George Muirhead

The Birds of Berwickshire

With Remarks on their Local Distribution Migration, and Habits, and also on the

Folklore, Proverbs, Popular Rhymes and Sayings Connected with them

THE
BIRDS OF BERWICKSHIRE

WITH REMARKS ON THEIR LOCAL DISTRIBUTION
MIGRATION, AND HABITS, AND ALSO ON THE
FOLK-LORE, PROVERBS, POPULAR RHYMES
AND SAYINGS CONNECTED WITH THEM

BY

GEORGE MUIRHEAD, F.R.S.E., F.Z.S.

MEMBER OF THE BRITISH ORNITHOLOGISTS' UNION, MEMBER OF
THE BERWICKSHIRE NATURALISTS' CLUB, ETC.

IN TWO VOLUMES

No more the screaming bittern, bellowing harsh,
To its dark bottom shakes the shuddering marsh.

VOL. I.

EDINBURGH: DAVID DOUGLAS
1889

> *Nature never did betray*
> *The heart that loved her; 'tis her privilege,*
> *Through all the years of this our life, to lead*
> *From joy to joy.*
> WORDSWORTH.

CONTENTS OF VOL. I.

		PAGE
INTRODUCTION,	.	xiii

PASSERES.

TURDIDÆ.

Turdus musicus.	Song Thrush,	1
,, *viscivorus.*	Missel Thrush,	8
,, *iliacus.*	Redwing, .	14
,, *pilaris.*	Fieldfare, .	18
,, *varius.*	White's Thrush, .	22
,, *merula.*	Blackbird, . .	25
,, *torquatus.*	Ring Ouzel, .	32

SYLVIIDÆ.

Saxicola œnanthe.	Wheatear, .	36
,, *rubetra.*	Whinchat, .	41
,, *rubicola.*	Stonechat, .	44
Phœnicura ruticilla.	Redstart, .	47
Erithaca rubecula.	Redbreast, .	50
Sylvia rufa.	Whitethroat, .	57
,, *atricapilla.*	Blackcap, . .	60
,, *salicaria.*	Garden Warbler, . .	63
Regulus cristatus.	Golden-Crested Wren, .	65
Phylloscopus collybita.	Chiffchaff, . .	68
,, *trochilus.*	Willow Wren,	71
,. *sibilatrix.*	Wood Wren, . .	73
Acrocephalus schœnobœnus.	Sedge Warbler, .	75
,, *nœvius.*	Grasshopper Warbler,	79
Accentor modularis.	Hedge Sparrow, .	83

CINCLIDÆ.

Cinclus aquaticus.	Dipper, ...	85

CONTENTS.

PASSERES—continued.

PARIDÆ.

		PAGE
Acredula caudata.	Long-Tailed Titmouse, .	89
Parus major.	Great Titmouse, . .	91
Parus ater.	Cole Titmouse, .	93
,, *palustris.*	Marsh Titmouse, .	95
,, *cœruleus.*	Blue Titmouse, .	97

SITTIDÆ.

Sitta cæsia.	Nuthatch,	100

TROGLODYTIDÆ.

Troglodytes parvulus.	Wren, .	102

MOTACILLIDÆ.

Motacilla lugubris.	Pied Wagtail,	108
,, *sulphurea.*	Grey Wagtail,	111
Anthus pratensis.	Meadow Pipit,	113
,, *trivialis.*	Tree Pipit, .	116
,, *obscurus.*	Rock Pipit, .	118

ORIOLIDÆ.

Oriolus galbula.	Golden Oriole,	120

LANIIDÆ.

Lanius excubitor.	Great Grey Shrike,	123
,, *collurio.*	Red-Backed Shrike,	127

AMPELIDÆ.

Ampelis garrulus.	Waxwing,	129

MUSCICAPIDÆ.

Muscicapa grisola.	Spotted Flycatcher,	131
,, *atricapilla.*	Pied Flycatcher, .	133

HIRUNDINIDÆ.

Hirundo rustica.	Swallow, .	135
Chelidon urbica.	Martin, . .	141
Cotile riparia.	Sand Martin,	144

CONTENTS. vii

PASSERES—continued.

CERTHIIDÆ.
		PAGE
Certhia familiaris.	Tree-Creeper,	146

FRINGILLIDÆ.
Carduelis elegans.	Goldfinch,	148
,, *spinus.*	Siskin,	152
Coccothraustes chloris.	Greenfinch,	154
Passer domesticus.	House Sparrow,	156
,, *montanus.*	Tree Sparrow,	160
Fringilla cœlebs.	Chaffinch,	162
,, *montifringilla.*	Brambling,	166
Linota cannabina.	Linnet,	168
,, *rufescens.*	Lesser Redpoll,	172
,, *flavirostris.*	Twite,	174
Pyrrhula europœa.	Bullfinch,	176
,, *enucleator.*	Pine Grosbeak,	178
Loxia curvirostra.	Crossbill,	179

EMBERIZIDÆ.
Emberiza miliaria.	Corn Bunting,	180
,, *citrinella.*	Yellow Bunting,	183
,, *schœniclus.*	Reed-Bunting,	186
Plectrophanes nivalis.	Snow-Bunting,	189

STURNIDÆ.
| *Sturnus vulgaris.* | Starling, | 192 |

CORVIDÆ.
Pyrrhocorax graculus.	Chough,	197
Garrulus glandularius.	Jay,	200
Pica rustica.	Magpie,	202
Corvus monedula.	Jackdaw,	207
,, *corone.*	Carrion Crow,	210
,, *cornix.*	Hooded Crow,	214
,, *frugilegus.*	Rook,	216
,, *corax.*	Raven,	240

ALAUDIDÆ.
| *Alauda arvensis.* | Skylark, | 250 |

CONTENTS.

PASSERES—continued.

		PAGE
CYPSELIDÆ.		
Cypselus apus.	Swift, .	256
CAPRIMULGIDÆ.		
Caprimulgus europæus.	Nightjar,	258
PICIDÆ.		
Dendrocopus major.	Great Spotted Woodpecker, .	261
Iynx torquilla.	Wryneck, ...	265
ALCEDINIDÆ.		
Alcedo ispida.	Kingfisher, .	267
UPUPIDÆ.		
Upupa epops.	Hoopoe,	271
CUCULIDÆ.		
Cuculus canorus.	Cuckoo, .	273

STRIGES.

STRIGIDÆ.		
Aluco flammeus.	Barn Owl, .	280
Asio otus.	Long-Eared Owl, .	287
,, *accipitrinus.*	Short-Eared Owl, .	287
Strix aluco.	Tawny Owl, .	292

ACCIPITRES.

FALCONIDÆ.		
Circus cyaneus.	Hen-Harrier, .	296
Buteo vulgaris.	Buzzard, . . .	302
,, *lagopus.*	Rough-Legged Buzzard,	305
Haliaetus albicilla.	Sea Eagle, . .	307
Accipiter nisus.	Sparrow-Hawk, .	311
Pernis apivorus.	Honey Buzzard, .	315

ILLUSTRATIONS

IN VOL. I.

BILLIE MIRE. By JOHN BLAIR, . *Title-page*	
MAP OF BERWICKSHIRE, . *facing page* xiii	
	PAGE
THE MAVIS'S NEST. By JOHN BLAIR, from Photograph by G. MUIRHEAD, . . .	7
PAXTON HOUSE. By JOHN BLAIR, .	13
SHEEP FEEDING ON TURNIPS IN WINTER AT NABDEAN. By JOHN BLAIR, .	17
HARDACRES, NEAR ECCLES. By JOHN BLAIR,	24
BLACKBIRD'S NEST. From Nature, by MRS. MUIRHEAD, Paxton,	31
DYE COTTAGE. By JOHN BLAIR, .	35
WEDDERLIE FARM-HOUSE. By JOHN BLAIR, .	40
ELLEMFORD BRIDGE. By JOHN BLAIR, from Photograph by G. MUIRHEAD, . . .	43
THIRLESTANE CASTLE. By JOHN BLAIR, from Photograph, .	49
AT SALTON, EAST-LOTHIAN. By JOHN BLAIR, from a Photograph by G. MUIRHEAD, October 1887, . .	56
WHITETHROAT'S NEST. From Nature, by MRS. MUIRHEAD,	59
BLACKCAP'S NEST. From Nature, by MRS. MUIRHEAD,	62

ILLUSTRATIONS.

	PAGE
DEAN AT MR. MUIRHEAD'S HOUSE, PAXTON. From Nature, by JOHN BLAIR,	64
GOLDEN-CRESTED WREN'S NEST. From Nature, by MRS. MUIRHEAD,	67
CHIFFCHAFF'S NEST. From Nature, by MRS. MUIRHEAD,	70
WILLOW WREN'S NEST. From Nature, by MRS. MUIRHEAD,	72
NABDEAN MILL POND. From Nature, by JOHN BLAIR,	78
GRASSHOPPER WARBLER'S NEST. By MRS. MUIRHEAD, from Nest sent from Cambridgeshire by A. H. EVANS,	82
THE YOUNG BIRDCATCHER. By JOHN BLAIR,	84
THE WHITADDER NEAR HUTTON MILL. By W. D. M'KAY, R.S.A.,	facing page 86
DIPPER'S NEST. From Nature, by MRS. MUIRHEAD,	88
PISTOL PLANTATION. From Nature, by JOHN BLAIR,	90
BEE. By MRS. MUIRHEAD,	92
SMALL STREAM NEAR GAMEKEEPER'S HOUSE, PAXTON. From Nature, by JOHN BLAIR,	94
PEASE MILL. By JOHN BLAIR,	96
BARN-DOOR IN WINTER, CAMPSTEAD, NEAR PAXTON. By JOHN BLAIR,	99
VIEW OF DUNS FROM THE EAST. By JOHN BLAIR,	101
WREN'S NEST. From Nature, by MRS. MUIRHEAD,	107
BROOMHOUSE. By JOHN BLAIR, from Photograph,	110
MEADOW PIPIT'S NEST. From Nature, y MRS. MUIRHEAD,	115
ROCKS AT BURNMOUTH. By JOHN BLAIR, from Photograph by H. EVANS,	119
COCKBURNSPATH VILLAGE. By JOHN BLAIR,	122
SWINTON KIRK. By JOHN BLAIR,	126

ILLUSTRATIONS.

	PAGE
STAINRIG. By JOHN BLAIR, .	130
SPOTTED FLYCATCHER'S NEST. From Nature, by MRS. MUIRHEAD,	132
SWALLOW FLYING. By JOHN BLAIR, .	140
SWALLOW CRAIG, NEAR OLDCAMBUS. From Nature, by JOHN BLAIR,	143
BLACKCASTLE RINGS. From Nature, by JOHN BLAIR, .	145
BOATHOUSE AT PAXTON. By JOHN BLAIR, .	147
OLD THIRLESTANE. By JOHN BLAIR,	151
EDINGTON MILL. By JOHN BLAIR,	153
STACKYARD IN WINTER—SPITAL MAINS. By JOHN BLAIR, .	155
EVELAW TOWER. By JOHN BLAIR, . .	159
HUTTON HALL. By JOHN BLAIR, from Sketch taken in 1844 by W. WILSON, Jun., kindly lent by DR. STUART of Chirnside, ...	161
CHAFFINCH'S NEST. From Nature, by MRS. MUIRHEAD,	165
VILLAGE OF OXTON. By JOHN BLAIR, . . *facing page*	167
LINNET'S NEST. From Nature, by MRS. MUIRHEAD, .	171
RHYMER'S TOWER, EARLSTON. By JOHN BLAIR, .	173
JUNCTION OF THE DYE AND THE WATCH. From Nature, by JOHN BLAIR, . .	175
BULLFINCH'S NEST. From Nature, by MRS. MUIRHEAD,	177
OLD MANSE AT ECCLES. By JOHN BLAIR,	178
AUCHENCROW VILLAGE. By JOHN BLAIR, .	182
TOAD IN YELLOW BUNTING'S NEST. By JOHN BLAIR, .	185
REED-BUNTING'S NEST. From Nature, by MRS. MUIRHEAD,	188
EDINGTON MAINS IN SNOW-STORM. By JOHN BLAIR, .	191
CORSBIE TOWER. By JOHN BLAIR, .	196
THE RETREAT. By JOHN BLAIR, from Photograph by G. MUIRHEAD, . .	201

ILLUSTRATIONS.

	PAGE
THE DOD MILL. By JOHN BLAIR, from Photograph by G. MUIRHEAD,	206
ST. HELEN'S KIRK. By JOHN BLAIR, from Photograph by G. MUIRHEAD,	209
GAMEKEEPER'S RAIL AT WEDDERLIE HOUSE. By JOHN BLAIR,	213
TWEED MILL. By JOHN BLAIR,	215
ROOK-SHOOTING AT PAXTON. By JOHN BLAIR,	239
FAST CASTLE. Original Etching by W. B. HOLE, R.S.A., *facing page*	243
ROCKS AT FAIRNEYSIDE, NEAR BURNMOUTH. By JOHN BLAIR, from Photograph by G. MUIRHEAD,	249
BILLIE CASTLE. By JOHN BLAIR,	255
DRYBURGH ABBEY. By JOHN BLAIR,	257
PENMANSHIEL. By JOHN BLAIR,	260
AYTON CASTLE. By JOHN BLAIR,	264
MILNE GRADEN HOUSE. By JOHN BLAIR,	266
THE TWEED AT PAXTON. From Original Etching by W. D. M'KAY, R.S.A. Engraved by G. AIKMAN, A.R.S.A., *facing page*	267
THE MERMAID'S WELL, NEAR THE DOD MILL. By JOHN BLAIR,	272
GREENKNOWE TOWER. By JOHN BLAIR,	279
LENNEL KIRKYARD. By JOHN BLAIR,	286
FINCHY SHIEL, ON THE TWEED AT PAXTON. By JOHN BLAIR,	295
BILLIE MIRE, LOOKING WEST. By JOHN BLAIR,	301
THE MEUCHEL STANE. By JOHN BLAIR,	304
EARNSHEUGH, NEAR ST. ABB'S HEAD. By JOHN BLAIR,	310
THE PECH STANE, NEAR BILLIE MAINS. By JOHN BLAIR,	314
PENMANSHIEL WOOD. By JOHN BLAIR,	316

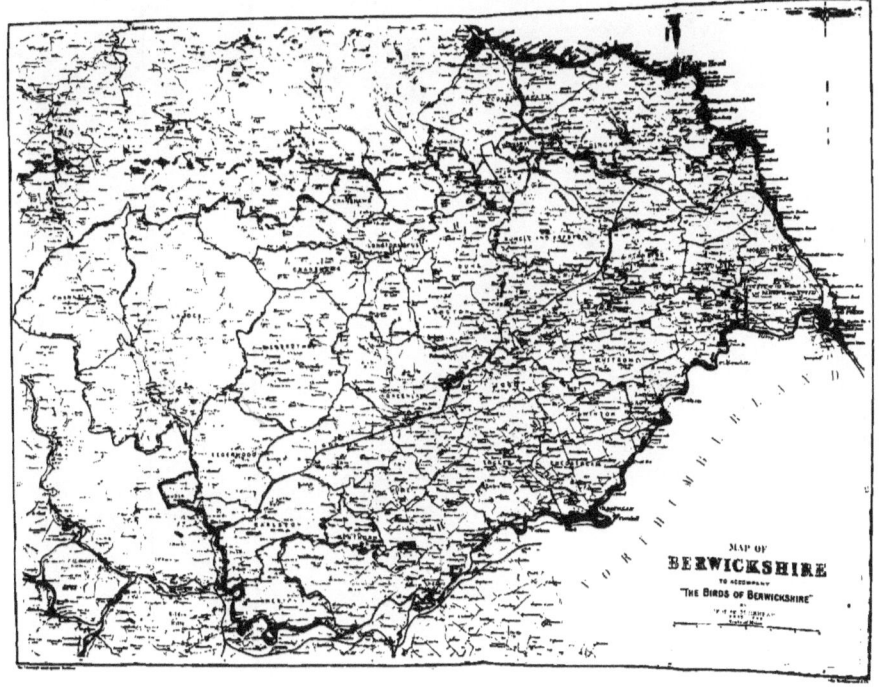

Middle Irwin   Beerboweyrimuse   Chwiedee

INTRODUCTION.

BERWICKSHIRE is the most south-easterly county in Scotland, and has, generally speaking, well-defined natural boundaries — the Lammermuir Hills on the north, the valley of the Leader on the west, the German Ocean on the north and east, and the river Tweed on the south.[1] It is oblong in shape—the greatest length from east to west being about 29 miles, and the utmost breadth from north to south nearly 21 miles. The area is 294,804 acres of land, 1557 acres of water, and 799 acres of foreshore—in all 464 square miles.[2]

The coast is bold and precipitous, and measures, exclusive of minor sinuosities, about 19 miles in length. It trends in a north-westerly direction from the eastern extremity of the parish of Mordington—on the march of the Liberties of Berwick — to St. Abb's Head, and thence westwards by Fast Castle to the mouth of Dunglass Burn on the boundary of East-Lothian. Along the seaboard a wide ridge of elevated ground stretches from Lamberton Moor on the east to the neighbourhood of Cockburnspath on the west, where it terminates in the Lammermuir Hills, which extend thence along the northern and western confines of the county to Channelkirk and

---

[1] According to the one-inch Ordnance Survey the limits of Berwickshire are as follows :—

Northern limits,	55° 56′ 50″
Southern limits, .	55° 34′
Eastern limits, .	2° 2′
Western limits, ...	2° 54′ 35″

[2] See *Ordnance Gazetteer of Scotland* (Edinburgh, 1886), vol. i. p. 152.

Lauder—a distance of about 27 miles. From the southern extremity of the latter parish the Leader forms the limit for about six miles, with the exception of a small part of its course at Chapel; and after watering the beautiful haughs of "Ercildoun and Cowdenknowes," it flows into the Tweed below Drygrange. The windings of this classical stream encircle Bemerside, Dryburgh, and Mertoun, and divide the county from Roxburgh along a course of ten miles until Makerstoun is reached. Here the line of demarcation becomes artificial, and after passing tortuously round Smailholm and Stitchel for about twenty miles, again strikes the Tweed near Birgham, which "fair river," flowing past Coldstream and "Norham's castled steep," separates Berwickshire from Northumberland for about eighteen miles, until it enters English ground at Paxton. At this point the march runs northwards for two miles along the Old Bound Road as far as Mordington, and thence for the same distance eastwards to the German Ocean.

Between the Lammermuir Hills and the Tweed a wide undulating plain, known as the Merse, extends to upwards of 100,000 acres. It is rich and highly cultivated, and is adorned with so many beautiful woods and plantations, that the general landscape of its central district assumes a sylvan aspect in summer, when viewed from some of the surrounding high grounds, such as those in the neighbourhood of Hardens, Foulden, or Mordington.

Describing the view from the heights above Chirnside, the late Rev. John Edgar of Hutton says: "It is doubtful whether in any district of Scotland such an extensive, rich, and well-cultivated rural panorama can be found, or whose external features bear so near a resemblance to some of the fertile plains of 'Merry England.' The Lammermuir

Hills are seen in all their sterile and heath-clad blackness, their rounded forms well defined, and placed, as it were, as a protection to the wide and cultivated plains which they enclose. Extending for many miles, the Merse appears reclining in calm repose, its surface decked with various objects of rural interest, and interspersed with trees, hedgerows, woods, rich pastures, and spacious and fertile cornfields. To the south-west is seen Home Castle, frowning from its elevated site on the plain beneath, and recalling the remembrance of feudal times. At a further distance, and nearly in the same direction, appear the Eildon Hills, towering to the clouds, while far to the south-east the huge masses of the Cheviot Mountains, rising in dim and dusky grandeur, arrest the eye of the spectator, and furnish a fine and imposing termination to the scene." [1]

Although the Merse is now so beautifully wooded and well cultivated, it was for many centuries previous to 1730 almost wholly destitute of trees, and little or no progress had been made in the enclosing or improvement of the ground. Here it may not be considered out of place to take a rapid glance at some of the more important changes which have modified its surface within the historical period.

When the Romans, in their conquest of Britain, reached this district, it is believed that they found the greater part of it covered with extensive forests and marshes, in which roamed the Wild Ox,[2] the Red Deer,[3]

---

[1] *New Statistical Account of Scotland*, 1845, vol. ii. (Berwickshire), p. 362.

[2] See "Notes on the Ancient Cattle of Scotland," by Dr. J. A. Smith, in *Proc. Soc. Antiq. of Scot.*, 1873, vol. ix. part II. p. 641.

[3] In olden times the Red Deer abounded in the Lammermuirs, where it has given its name to North and South Hartlaw, also to the farms of Hartside and Hindsidehill. Its remains have been found at Whitrig Bog in the parish of Mertoun; Whiteburn in the parish of Westruther; and Kimmerghame in the parish of Edrom. For some interesting notes by Mr. Hardy on the Red Deer in Berwickshire, see *Hist. Ber. Nat. Club*, vol. iv. pp. 214-217; also Ridpath's *Border History*, p. 541.

the Roe,[1] the Wolf,[2] and the Wild Boar,[3] whilst on many of the inland waters the Beaver[4] still constructed his dam. The invaders are credited with burning immense tracts of these woods to clear the ground, and to enable them to carry on a more successful warfare with the native Britons—the Otadini—whom they could not follow into their sylvan fastnesses, as well as to drive off the numerous Wolves and Wild Boars with which the country was infested.

After the final retirement of the Romans in 426 A.D., the region was successively ravaged by the Saxons and the Danes, who seem to have completed the destruction which their predecessors had begun, for in the time of William the Lion (1165-1214) the woods had almost wholly disappeared, and those which remained were confined to the sheltered valleys and the banks of the rivers.[5]

These great changes in the face of the country would doubtless be followed by corresponding alterations in the fauna of the district; the various species, to whose welfare

---

[1] This beautiful animal was found in the country until a comparatively recent period. In a lease of the "landys of Brockholl, Hernode, and Denewod," granted to Thomas Atkynson in 1429-30, the "venyson" is reserved.—(*Priory of Coldingham, Surtees Soc.*, p. 105.) Several places in the Lammermuirs have derived their names from the Roe, amongst which may be mentioned Raecleughhead, in the parish of Langton, and Rawburn (Roeburn), in the parish of Cranshaws.

[2] See "History of the Wolf in Scotland," by Mr. James Hardy; *Hist. Ber. Nat. Club*, vol. iv. pp. 289-292.

[3] See the old *Stat. Account of Scot.*, vol. v. p. 88, and vol. vi. pp. 322-23— also the *Swintons of that Ilk*, by Mr. Campbell-Swinton of Kimmerghame, 1883, pp. 1, 2. That the Wild Boar was found in the neighbourhood of Swinton in olden times there can be little doubt. A field adjoining the village is known by the name of "The Sow Mire" to this day.

[4] *Castor europæus*. There is indubitable evidence of the existence of the Beaver in Berwickshire in ancient times, for, in draining Middlestots Bog on the estate of Kimmerghame, in October 1818, its remains were found imbedded partly in the marsh and partly in a layer of peat-moss. See Neill in *Edin. Phil. Journ.*, i. p. 84; and in *Wern.*, iii. p. 216; also Milne in *Essay on the Geology of Berwickshire*, p. 229 (Dr. Johnston's MS. Notes).

[5] Carr's *History of Coldingham Priory*, p. 27.

the forests were essential, diminishing, and the others increasing. As civilisation advanced, and the inhabitants became more numerous, many wild animals, such as the Red Deer, the Wolf, and the Wild Boar, were driven away or extirpated; sheep or cattle now browsed on all the drier parts of the ground, while as yet bogs and morasses, interspersed with sheets of water, covered a considerable portion of the surface. It is interesting to find that Hector Boece —the first British author who gave an account of the Great Bustard (*Otis tarda*), a bird which is only found on extensive and open plains—wrote of it in 1526 as then inhabiting the Merse. The date of its disappearance is unknown, nor has it left a memorial of its haunt in the name of any place in the locality.[1]

One of the chief features in the ornithology of the district in olden times would be the wild fowl, for in the numerous shallow lochs, like those of Legerwood and Bemerside, they doubtless abounded in every variety, from the diminutive Teal to the stately Wild Swan; whilst the extensive and almost impenetrable bogs, such as Billie Mire, afforded secure retreats in which the resident species could lay their eggs, and rear their young in safety. At nightfall, in the spring and early summer months, the hollow boom of the Bittern, or "Bull o' the Bog," was heard resounding through the marshes, mingled with the weird cries of innumerable water-fowl of various kinds. In the dusk of the evening great flocks of Wild Ducks were seen winging their way to their adjacent feeding-grounds; while it is probable that the Crane [2] (*Grus cinerea*) was also found here in those days, for

---

[1] A specimen of the Great Bustard (*Otis tarda*)—a female—was shot at Fenham Flats, near Holy Island, on 2nd January 1871. It is now preserved in the Berwick Museum.

[2] An example of this interesting bird was shot at Threepwood, on the confines of Berwickshire, in the neighbourhood of Lauder, some years ago.—*Hist. Ber. Nat. Club*, vol. vii. p. 305.

it appears to have been then a well-known bird in Scotland,[1] and is thus referred to by Gavin Douglas in his translation of Virgil :—

> Palamedes birdis crowpand in the sky,
> Fleand on randoun, schapin lyk ane Y,
> And as an trumpit rang thare vocis soun,
> Quhais cryis bene pronosticacioun
> Of wyndy blastis and ventosities.[2]

According to Sir David Lindsay it was served at a grand hunting entertainment, given by the Earl of Athol to James V. of Scotland and the Queen-Mother, in Glen Tilt;[3] and in 1551 the Scottish Parliament fixed the price of the "Cran" at "five shillinges."[4]

Wild-fowl seem to have been the chief game at which Falcons were flown, and many severe laws were made to preserve them for this sport. To prevent them from being killed with the gun, which was then beginning to come into use for fowling, Queen Mary even punished with death any of her lieges who, "of quhatsumever degree he bee of, take upon hande to schutte at wilde-fowls with half-hag, culvering, or pistolet."[5]

In the Accounts of the Lord High Treasurer of Scotland there are frequent references to King James IV. visiting Home Castle in the Merse, with his falconers, where Lord Home doubtless would show his royal guest abundance of fowl, including Wild Geese, Herons, and Bitterns, to try the mettle of the Hawks.

Partridges were found in small numbers in the neigh-

---

[1] Lesley, in his description of Scotland, says, "Grues plurimi, sicut et ardeæ : olores autem, quorum apud Anglos magnus est proventus, pauciores."—*De origine et rebus gestis Scotorum*, ed. 1578, p. 25.
[2] *Chronicles of Scottish Poetry*, Sibbald, vol. i. p. 431.
[3] Yarrell's *British Birds*, 4th ed., vol. iii. p. 179.
[4] *The Laws and Acts of Parliament made by the Kings and Queen of Scotland.* Collected by Thomas Murray of Glendook, Knight. Edinburgh, 1681.
[5] *Ibid.*

bourhood of the scanty patches of corn, Godscroft recording in 1611 that Sir George Home of Wedderburn took them with Falcons in the latter part of the previous century.[1] The Pheasant had not then been introduced.

Birds of prey abounded—the White-Tailed Eagle soared in the sky, the Peregrine, the Merlin, and the Sparrow-Hawk swept the hills and the plains, whilst the Buzzard and the Hen-Harrier hunted the fields and marshes. The hoarse cry of the Raven was ever present, and the chattering of the Magpie and the Jay was constantly heard in the wooded valleys.

In the early part of the eighteenth century the surface of the Merse, which had remained bare for ages, began to be transformed once more. The proprietors enclosed, planted, and drained their lands, which, under the skilful management of the tenants, now bore luxuriant crops of grain, where previously the rush flourished and the reed was cut for thatching the cottage of the peasant. The lochs, swamps, and bogs gradually disappeared until only a few of the largest remained. Amongst these was Billie Mire, a deep and extensive morass which lay along a narrow valley from the vicinity of Ayton to the neighbourhood of Chirnside, a distance of about five miles. Besides being of great interest from an ornithological point of view as the former haunt of the Hen-Harrier, the Bittern, and innumerable wild-fowl, it was famous in border history for having formed an impassable barrier to the English invaders, and for a treaty between the two kingdoms concluded on its banks in 1386.[2] The drainage of this great bog, which for ages had been the resort of myriads of water-fowl, was begun in 1801, but for some years later the Harrier continued to nest in the sedges, and the "Bull o' the Mire" boomed from the reeds. About

---

[1] *History of the Homes of Wedderburn*, by David Home of Godscroft. 1611.
[2] See Carr's *History of Coldingham Priory*, pp. 31, 32.

1830 the work was finally completed, and the spongy swamps—the ancient home of the Black-Headed Gull, the Snipe, and the Water-Hen—were finally laid dry. Whilst the haunts of the Wild Duck were thus destroyed, vast flocks of Wild Geese still continued to visit every part of the Merse during the autumn, winter, and early spring months, where they fed on the young grass and wheat-fields.

The climate of the district had greatly improved, and ague, which had hitherto tormented the rural population, was gradually banished. Woods now formed a striking feature in the landscape, and afforded abundant shelter to the Blackbird, the Mavis, and the smaller warblers, whose joyous songs were heard where formerly only the melancholy cry of the Plover disturbed the silence of the plain. The Pheasant, which had been introduced about 1815 at Thirlestane Castle, and a few years later at Spottiswoode and The Hirsel, began to attract the attention of the landed proprietors, the plantations affording sufficient cover for this beautiful bird. About 1840 the preservation of game, which hitherto had received only partial attention, seems to have become general. Partridges increased to a great extent, owing to the abundance of food supplied by the enclosed and well-cultivated fields and the destruction of birds of prey, such as the Hen-Harrier and the Sparrow-Hawk, together with their four-footed enemies.

After the middle of the present century the Missel Thrush became comparatively plentiful; and the Starling, as well as the Wood Pigeon, multiplied greatly. In many of the rocky precipices on the banks of the Tweed, Whitadder, and Blackadder, the Barn Owl still nested and reared its young, although it has now (1889) almost entirely disappeared, whilst its tawny congener is heard hooting in every plantation. In the summer of 1877 the Stock Dove (*Columba œnas*) appeared at Paxton, and is now rapidly extending along the wooded banks of the streams, and in other suitable localities. Rooks, Jackdaws, and House Sparrows have increased in nearly every parish. In June 1888, Pallas' Sand Grouse (*Syrrhaptes paradoxus*), a new visitor to the district, was found at Foulden West Mains, and shortly afterwards in the neighbourhood of Abbey St. Bathans and Quixwood.

Having thus briefly surveyed the level region of the Merse, let us now turn to the upland districts of the county.

The Lammermuirs consist of a broad range of rounded but lofty hills,[1] rising at Seenes Law, in the parish of Lauder, to 1683 feet above the level of the sea. They are intersected by numerous valleys, through many of which small streams, fed by sykes and rills from the adjacent bogs, gurgle over their rocky beds, and form the sources of the Whitadder, Blackadder, Leader, Eden, Dye, and Eye, which water the lower grounds of the county.

On the outskirts of the hills cultivation has extended, cattle and sheep now browse there on green pastures, and the yellow grain waves in the autumn sun; but, higher up, the purple heather still perfumes the air. We are there in the home of the Red Grouse, the Black Cock, the Curlew, and the Plover; and Nature can be seen in much the same garb as she appeared to the hunter of old as he pursued the Red Deer, the Wolf, and the Wild Boar in their native

---

[1] Eighteen of the highest summits, with their respective altitudes above sea-level, are Tarf Law (1248 feet), Dun Law (1292), Black Hill (1299), Byrecleugh Ridge (1335), Lamb Rig (1339), Wether Law (1379), Hog Hill (1395), South Hart Law (1437), Wedder Law (1460), Ninecairn Edge (1479), Waddels Cairn (1490), Meikle Law (1531), North Hart Law (1578), Wedderlairs (1593), Hunt Law (1625), Willie's Law (1626), Crib Law (1670), and Seenes Law (1683).—*Ordnance Gazetteer of Scotland*, 1886.

haunts. In ancient times the whole of the country on the outskirts of the hills was covered with heather, which also stretched far into the level plains of the Merse. The plough has invaded the outlying domains of the Grouse, but little change has taken place in the character or surface of the higher grounds, and, consequently—with the exception of some of the larger birds of prey, such as the Hen-Harrier and the Raven, which have been exterminated—the species of the winged inhabitants of the district remain much the same as of yore.

In the early years of this century, before the days of game preservation, Grouse appear to have been very scarce in the Lammermuirs, for we find from the diary of a Berwickshire sportsman, who, in August and September 1812-14, frequently visited Longformacus, Monynut, and Bedshiel, to shoot Moor-fowl, that he did not, upon any occasion, see more than fourteen, or get above one shot at the birds in a day, on account of their wildness.

As a contrast to this it may be mentioned that in 1872 upwards of 700 brace were bagged in a fortnight on the Byrecleugh Moors; and on the 12th, 13th, and 14th of August 1886, 386 brace were killed in the neighbourhood of Carfrae Mill. There can be little doubt that this vast increase has been caused by the strict preservation of the ground from poachers, the destruction of all the winged and four-footed enemies of the game, and the judicious burning of the heather, as well as the killing down of the old birds, by driving them over batteries in which the sportsmen are concealed.

Whilst Grouse have multiplied so marvellously, Blackgame, probably owing to the drainage of many of their favourite haunts, do not seem to have advanced in numbers, for they are not plentiful in any locality, being apparently more numerous in the vicinity of Spottiswoode than elsewhere.[1] A few years ago several took up their quarters on Lamberton Moor, where they now breed again, this being an old home of the bird previous to 1870, when the last remaining Grey Hen was seen.

In former days the Dotterel (*Charadrius morinellus*) visited the high grounds along the sea coast and many localities in the Lammermuirs in great flocks, on its spring migration in April and May,—the uplands about Mellerstain, Legerwood, Abbey St. Bathans, Redheugh, Coldingham, and Lamberton being favourite haunts,—and thither the local sportsmen resorted to shoot them for the table. It is now, however, seen only in small numbers, and at irregular intervals, chiefly on the moors about Redheugh, Dowlaw, and Lamberton.

In autumn and spring some parts of the hills are visited by immense flocks of Wild Geese, the Hule Moss on Greenlaw Moor, Gibb's Cross at Wedderlie, and the Braid Bog near Penmanshiel, being some of their favourite resorts.

The wild whistling notes of the Curlew and the Golden Plover, as well as the scream of the Lapwing, are heard in summer all over the moors; for here they have their nests, and rear their young. The Snipe frequents the bogs, where it drums in spring, and has its home among the rushes. At the same season the song of the Ring Ouzel cheers the weary shepherd, whilst the Wheatear and the Whinchat flit chirping from stone to stone. The Sky Lark rises from the dewy grass to salute the morn; the timid Titling peeps in the heather; and, in the neighbouring vale, the voice of the wandering Cuckoo falls upon the ear. When the hillsides are glowing with the golden bloom of the whin, the Linnet builds her bower in the prickly fortress, the Twite nestles in the heath; and the fragrant larches, here and

---

[1] See Mr. Harvie-Brown's remarks on the "Decrease of Black-game," in his interesting work on *The Capercaillie in Scotland*; Edinburgh, 1878, pp. 120-126.

there fringing the sides of the rocky streams, afford shelter to the Willow Wren, which gladdens the solitary angler with its song.

In winter the region is very desolate, wandering flocks of Snow Buntings being the only small birds which frequent the gloomy hills at this dreary season of the year.

The Lammermuirs slope down to the sea in the neighbourhood of Cockburnspath, where the scenery changes completely, for we now look out upon the wide expanse of the German Ocean, and hear the sound of the restless waves dashing against the rock-bound shore.

The whole of the coast of Berwickshire is bold and majestic, with numerous lofty precipices, which rise at Earnsheugh to the height of five hundred feet above the surging waters which wash their base, being the highest cliffs on the eastern sea-board of Great Britain. Here and there along this iron-bound shore great craggy rocks, towering isolated stacks, and low skerries stand out as bulwarks against the inroads of the ocean, whilst at intervals may be seen dark caverns hollowed out by the ceaseless action of the tides, in which the Rock Dove rears its young in safety amid the spray of waves which never rest.

On yonder point, where the shelving reefs are being rapidly covered by the advancing tide, we see a small group of Herons patiently waiting for their finny prey; and on a dark crag a little further from the beach the swarthy Cormorant sits in undisturbed repose. Flocks of Mallards float in security on the surface of the heaving waters, the Golden Eye dives at intervals beneath the surf, whilst the Eider displays his parti-coloured plumage to his sober-suited mate.

When the sky lowers, and the turbulent ocean, rendered furious by the rising winds, wreathes the sea-girt rocks with foam, and drives the white spray high above the beetling cliffs, then the note of the Northern Hareld strikes the ear from the midst of the wild waters; the wail of the Great Northern Diver—the "Herdsman of the Deep"—is heard above the booming of the billows, and the Storm Petrel is seen riding upon the tempest.

In spring the rocky coast is enlivened by the shrill whistle of the Curlew, the Redshank, and the Oystercatcher; far to seaward the eye is attracted by the white plumage of the Gannet returning to its home on the Bass, whilst nearer the shore the Herring Gull wheels in airy flight.

Proceeding eastwards, the ruins of Fast Castle—the "Wolf's Crag" of the *Bride of Lammermoor*, and, of old, the fastness of one of the Gowrie conspirators[1]—are seen mouldering on a lofty sea-girt rock; the harsh scream of the Peregrine, as the bird flies from its historical eyrie in the adjoining cliff, and the weird cries of the sea-fowl on their rocky citadels, being in singular unison with the wildness and desolation of the scene.

We now pass the Brander Cove—a resort of the Green Cormorant; Earnsheugh, a name suggestive of the Sea Eagle, where the Common Gull is *said* to have nested in former times; Ravensheugh, of old the habitation of the bird of ill omen; and arrive at Petticowick, the ancient but now deserted home of the Chough.

Leaving this romantic and picturesque little harbour, we row close round the rugged base of West Hurker, as well as the lofty isolated rocks of Flot Carr and Skelly, reaching the Rampart—in bygone days the haunt of the Kittiwake—before the full grandeur and magnificence of the scenery burst upon the view. The stupendous precipices of Fowl Carr, and the opposing perpendicular cliffs

---

[1] Logan of Restalrig. See Carr's *History of Coldingham Priory*, pp. 90-95, and 202-213.

covered with countless sea-fowl, now rise before the astonished eye, while the mid-air is darkened by the wings, and the ear deafened by the screams, of innumerable Guillemots and other birds descending from the rocky ledges, to which they come—

> At Nature's summons their aërial state
> Annual to found, and in bold voyage steer
> O'er this wide ocean, through yon pathless sky,
> One certain point to one appointed shore,
> By Heav'n's directive spirit here to raise
> Their temporary realm, and form secure,
> Where food awaits them copious from the wave,
> And shelter from the rock, their nuptial leagues;
> Each tribe apart, and all on tasks of love,
> To hatch the pregnant egg, to rear and guard
> Their helpless infants, piously intent.
> MALLET.

THE BIRDS OF BERWICKSHIRE

THE SONG THRUSH.

THRUSH, COMMON THRUSH, GARDEN THRUSH, WOOD THRUSH,
THROSTLE, MAVIS.

Turdus musicus.

The Mavie, The Mavis.

Within a thick and spreading hawthorn bush,
That overhung a molehill large and round,
I heard from morn to morn a merry Thrush
Sing hymns to sunrise, and I drank the sound
With joy ; and, often an intruding guest,
I watched her secret toils from day to day,—
How true she warped the moss to form a nest,
And modelled it within with wood and clay ;
And by and by, like heathbells gilt with dew,
There lay her shining eggs, as bright as flowers,
Ink-spotted-over shells of greeny blue ;
And there I witnessed in the sunny hours,
A brood of nature's minstrels chirp and fly,
Glad as that sunshine and the laughing sky.

CLARE, *The Thrush's Nest.*

THE delightful song of this favourite bird is associated with the return of spring in all its gladsomeness, with its wealth of wild-flowers, and trees bursting into leaf; for, in that joyful season, the "Throstle with his note so true" welcomes the morning,[1] and—

> The Mavis wild, wi' many a note,
> Sings drowsy day to rest.

---

[1] An ancient song, quoted by Gawin Douglas and Dunbar, and which apparently must have been known before 1500, begins :—

"Hay, now the day dawis, The Thissell Cok cryis
The jolie Cok crawis,   On lovers wha lyis,
Now shrouds the shawis,   Now skaillis the skyis,
Throu natur anone ;   The nicht is near gone."
            LEYDEN's Notes to *The Complaynt of Scotland.*

Amongst the chorus of birds which may be heard in our woods and groves on fine evenings in the pleasant month of May,[1] the melodious strains of the Thrush constantly fall upon the enraptured ear. "Listen!" says Macgillivray, "to the clear, loud notes that, in the softened sunshine, are poured forth in such wild melody. What do they resemble?—

> Dear, dear, dear,
>   Is the rocky glen;
> Far away, far away, far away
>   The haunts of men.
> Here shall we dwell in love,
> With the lark and the dove.
> Cuckoo and corn-rail;
> Feast on the banded snail,
>   Worm, and gilded fly:
> Drink of the crystal rill,
> Winding adown the hill,
>   Never to dry.
> With glee, with glee, with glee,
>   Cheer up, cheer up, cheer up; here
> Nothing to harm us; then sing merrily,
> Sing to the loved ones whose nest is near.
>   Qui, qui, qui, kween, quip.
>   Tiurru, tiurru, chipiwi,
>   Too-tee, too-tee, chin choo,
>   Chirri, chirri, chooee,
>   Quin, qui, qui!"

The Thrush is generally considered to be the best of our larger singing birds, and its song continues from early spring[2] until summer is well advanced. Sir Walter Scott,

---

[1] The Scottish poet, Alexander Scott, who wrote in the time of Queen Mary, says in his "Gratulation to the moneth of May":—

> In May the pleasant spray upsprings,
> In May the mirthful Mavcis sings.
>   *Chron. Scot. Poet.*, SIBBALD, iii. p. 161.

[2] A Mavis was heard singing at Paxton as early in the season as 23d January 1888; one was also heard as late as 11th November 1887.

in *Marmion*, alludes to the earliness of its song when he says—

> To dear Saint Valentine no Thrush
> Sings livelier from a spring-tide bush :

and Grahame, in his *Birds of Scotland*, has given us some beautiful lines on the variety and mellowness of its notes—

> The Thrush's song
> Is varied as his plumes ; and as his plumes
> Blend beauteous, each with each, so run his notes
> Smoothly, with many a happy rise and fall.
> . . . . . . So loud and soft,
> And high and low, all in his notes combine,
> In alternation sweet, to charm the ear.
> Full earlier than the blackbird he begins
> His vernal strain.

This species does not usually congregate in flocks like the Missel Thrush, Fieldfare, and Redwing, but is generally found scattered all over our district, in gardens, shrubberies, woods, and deans. Small flocks have, however, sometimes been observed in the uplands towards the end of autumn.[1] It is to a great extent migratory in Berwickshire ; but sufficient numbers of our home-bred birds remain in the county all the year round, in ordinary seasons, to cause a general belief that it is a permanent resident. It is regularly seen in large flocks in autumn, while on migration from the north of Europe, passing the lighthouses on the coasts of England and Scotland, including those at the Isle of May in the Firth of Forth, and the Farne Islands on the coast of Northumberland. On its return journey to the north in spring, it is again observed at the lighthouses above mentioned, but in fewer numbers ;[2] and in

---

[1] Mr. Hardy, writing in 1875, says :—" At Oldcambus, Berwickshire, on Dec. 2nd, the local Thrushes, during a frost, formed a small scattered flock in a field near a plantation. In former seasons I have frequently observed, at the close of autumn, Thrushes flying in flocks on the moors above Redheugh. They took refuge at night in the furze bushes."—*Hist. Ber. Nat. Club.* vol. vii. p. 511.

[2] Mr. John Cordeaux, author of the *Birds of the Humber District*, and Reporter on the Migration of Birds on the east coast of England, remarks, with regard to

some springs, such as those which followed the severe winters of 1879-80 and 1880-81, no Thrushes were observed at any of the lighthouses on the east coast of Scotland.[1]

A portion of the migrants which come across the North Sea in autumn make their way to Berwickshire, and swell the ranks of our resident birds. During their sojourn here they are often seen in great numbers amongst turnips, by partridge-shooters in September and October, where they appear to be engaged in searching for worms and insects. They rise singly out of the drills from under the leaves, as the sportsman advances, and, after flying a short distance, alight again. Having remained a few weeks with us, and having been joined by a great proportion of our native Thrushes, they proceed on migration to the south, like—

> The birds that bring summer, and fly when 'tis o'er.
> MOORE.

---

spring migration, that the reason why fewer birds are then observed at the lighthouses, as compared with autumn, is probably due to the fact that, in spring, birds migrate, with rare exceptions, at night ; and, as the weather is then finer and the nights clearer and shorter, they do not run their heads so much against the lanterns of the lighthouses and lightships. The spring migration is also carried on more leisurely—migration proceeding by easy stages northward.—*Report on Migration of Birds:* Spring and Autumn of 1880, p. 31.

[1] *Autumn* 1879.—Song Thrushes were observed on migration, at lighthouses on east coast of Scotland, from 16th Sep. to 25th Oct. ; at Farne Islands, from 15th Oct. to 5th Dec. *Spring* 1880.—No records from east coast of Scotland, or from Farnes, of any Song Thrushes being seen. *Autumn* 1880.—No records from east coast of Scotland. At Farnes, seen from 29th Oct. to 26th Nov. *Spring* 1881.—No records from east coast of Scotland ; at Farnes, on 2nd May, some Song Thrushes seen. *Autumn* 1881.—Records from east coast of Scotland, including Isle of May, from 22nd Sep. to 8th Dec. ; at Farnes, from 18th to 23rd Oct. *Spring* 1882.—No records from east coast of Scotland or from Farnes. *Autumn* 1882.—Records from east coast of Scotland, including Isle of May—Earliest at Isle of May, 22nd Sep. ; at Farnes, 27th Sep. to 15th Jan. 1883. *Spring* 1883.—East coast of Scotland—From 2nd Feb. to 10th May, at Bell Rock ; on 2nd March, at Isle of May ; at Farnes, from 20th Jan. to 10th May. *Autumn* 1883.—East coast of Scotland—Two seen at Isle of May on 28th Aug. (the earliest yet recorded in Scotland in autumn) and vast flocks from 30th Oct. to 3rd Nov. ; at Farnes, 21st Sep. to 2nd Jan. 1884. *Spring* 1884.—East coast of Scotland—At Isle of May, 2nd Feb. to 30th April ; at Farnes, 6th March to 1st May. *Autumn* 1884.—East coast of Scotland.—At Isle of May, 11th Sep. to 22nd Dec. ; at Farnes, 18th Oct. to 12th Nov.—Extracts from *Reports on the Migration of Birds*, from 1879 to 1884.

The terrible severity of the winters of 1878-79, 1879-80, and 1880-81, so thinned the ranks of our Berwickshire Song Thrushes, that several years elapsed before they recovered anything like their former numbers;[1] and their scarcity in our woodlands and shrubberies was suggestive of the following beautiful verses by Barry Cornwall—

> Whither hath the Wood Thrush flown
> From our greenwood bowers?
> Wherefore builds he not again
> Where the whitethorn flowers?
>
> Bid him come! for on his wings
> The sunny year he bringeth,
> And the heart unlocks its springs
> Wheresoe'er he singeth.

It would appear that the great deficiency of these birds in Berwickshire, from 1879 to 1882, was caused by the severity of the winters and springs of those years forcing them to migrate to the south, where they nearly all remained during the breeding seasons. Mr. Harvie-Brown of Dunipace, the well-known ornithologist, referring to their rarity in Scotland in the summer of 1879, says :—" Whilst the scarcity occurred in Scotland, unusual numbers of Thrushes hatched in the southern counties of England, being no doubt a large proportion of our Scotch population of ordinary seasons."[2]

---

[1] Mr. Hardy, writing on 16th April 1879, says :—" I went to examine some miles of wood on the banks of the Pease Burn. Birds were almost absent, and only *one* Thrush was heard."—*Hist. Ber. Nat. Club.* vol. ix. p. 129. Mr. John Ferguson, Duns, on 2nd May 1879, observes :—" I seldom see a Thrush, and several observers have remarked to me that they seem almost extinct."—*Ibid.* vol. ix. p. 136. Mr. Scott, gardener, Ladykirk House, 11th May 1879 :— "Thrushes have been reduced very much."—*Ibid.* vol. ix. p. 138. Mr. W. Cunningham, Rosybank, Coldstream :—" There is scarcely a Thrush left."—*Ibid.* vol. ix. p. 140. Mr. Peter Loney, Marchmont, 7th May 1879 :—" We have no Thrushes yet."—*Ibid.* vol. ix. p. 140. Dr. Stuart, Chirnside, reports :—" Winter —1881-82 has almost exterminated the Common Thrush near Chirnside."—*Ibid.* vol. ix. p. 554. Mr. Robert Renton, Fans, Earlston, notes, 18th May 1883 :— "Two nests of Thrushes. This bird has not been seen here for the last two years."—*Ibid.* vol. x. p. 572.

[2] *Proceedings of the Nat. Hist. Soc. of Glasgow*, 30th Sep. 1879, p. 138.

This species feeds on snails, earthworms, insects, hips, fruit, and various berries, such as those of the yew, holly, and mountain-ash or rowan.

I have sometimes noticed it breaking the shells of snails which it finds by the sides of walls, garden palings, and such places; and have observed that it seizes the shell with its bill, and strikes it against a stone until it is broken, when it devours the contents.

Yarrell says that "in the vine countries of Europe it feasts luxuriantly during autumn on ripe grapes, and in many parts of the Continent it is in great request for the table at that time, from the excellent condition and flavour which abundance of this food imparts to its flesh;" and he adds that the beginning of the Drosselzug or Chasse aux Grives is regarded in many places nearly as the 12th of August and the 1st of September are with us. Phil Robinson, in the *Poets' Birds*, alluding to the use of song-birds for the table, remarks:—"Was ever the sweet myrtle so wickedly abused as in Sardinia, where they use it as a pickle for Thrushes?"

The Song Thrush begins to pair in March, and towards the end of that month, or the beginning of April—

> . . . . his song a partner for him gains;
> And in the hazel bush, or sloe, is formed
> The habitation of the wedded pair.
> GRAHAME, *Birds of Scotland*.

The nest is often built in a holly or yew in a garden or shrubbery; while in plantations it is frequently placed in a young spruce fir. It is composed externally of small sticks, roots, grass, and moss, and is lined with a thin layer of mud, cow-dung, or rotten wood.

It has been observed that the cock takes his share in incubating, and often feeds the hen while she is sitting. The eggs, which are from four to six in number, and of a beautiful shining greenish-blue colour, with black spots, are about thirteen days in hatching. Two broods are usually reared in the season.

The Mavis appears to have been a favourite cage-bird in Scotland long ago, for we find in the *Accounts of the Lord High Treasurer of Scotland*, 1473-98, the following entry in the reign of King James IV., under date 1497 :—

"Item, the xxvij day of Maij, gevin to Liale the falconare, quhen he brocht the King the Mavis fra Striuelin, be the Kingis command, . . . ix s."

THE MISSEL THRUSH.

MISTLETOE THRUSH, MISTLE THRUSH, GREY THRUSH, SCREECH THRUSH, SCREAMING MAVIS, THROSTLE COCK, STORM COCK.

Turdus viscivorus.

The Feltie, The Big Mavis, The Feltifleer.

> *While thou! the leader of the band,*
> *Fearless salut'st the opening year,*
> *Nor stay'st till blow the breezes bland,*
> *That bid the tender leaves appear;*
> *But on some towering elm or pine,*
> *Waving aloft thy dauntless wing,*
> *Thou joy'st thy love-notes wild to sing,*
> *Impatient of St. Valentine!*
>   C. SMITH, *To the Missel Thrush.*

IN the early years of this century the Missel Thrush seems to have been rarely seen in Berwickshire. Dr. Thomson, who drew up his report on the parish of Eccles for the *New Statistical Account of Scotland* in 1834,[1] says that it was very uncommon in that parish until within the three previous years; and Mr. John Wilson, late of Edington Mains, records that it was very scarce in his boyhood.[2] Writing on the 29th of May 1837, Mr. Hardy, Oldcambus, mentions it as a new settler in that neighbourhood, and, in 1845, he noted it as being found in Bunkle Woods.[3] Dr. Stuart, Chirnside, states that it was seldom observed when he came to reside

---

[1] *New Statistical Account of Scotland*, vol. ii., Berwickshire, p. 53.
[2] *Hist. Ber. Nat. Club*, vol. vi. p. 399.
[3] Mr. Hardy's *MS. Notes.*

in the county in 1847.[1] Mr. Gray remarks in 1871, that there has been a gradual increase in its numbers throughout Scotland during the last thirty years, and that, so recently as 1830, it was rather an unusual circumstance to find it breeding in any locality north of the Tweed.[2]

The Missel Thrush is now comparatively common in Berwickshire, where it is a partial migrant, and may be seen in many of our grass fields in small flocks during the autumn and winter months. It frequents woods and plantations in spring and summer.

The great increase of this bird in the county, within the last sixty years, is probably the result of various causes, which would likewise affect the whole of the south of Scotland. There can be little doubt that the destruction of Magpies and birds of prey for the preservation of game has promoted the increase of many of our smaller birds, and it is possible that this may partly account for the extension of the range of the Missel Thrush northwards from its former southern breeding quarters. Where Magpies abounded— which they did in many parts of the country before game was strictly preserved—the nest of this Thrush, being usually placed in an exposed situation in plantations, would be particularly liable to be robbed by them, notwithstanding the vigilance of the old birds, and their daring attacks upon all intruders. As Magpies became scarce, the numbers of nests which escaped being robbed would become larger, and the young, being allowed to fly, would gradually occupy all the vacant space available in their neighbourhood, and extension northwards would naturally follow. This extension may likewise have been influenced by causes affecting the migration of the bird, such as favourable winds or otherwise; for, as pointed

---

[1] *Hist. Ber. Nat. Club*, vol. v. p. 442.
[2] *Birds of the West of Scotland*, p. 73.

out by Mr. Harvie-Brown, these have a considerable effect on the distribution of migrants.[1] Besides these two probable causes of the increase of this species in the county within the last sixty years, it must be kept in view that the ground under plantations, which provide shelter for the birds, has vastly extended within that period.

It sometimes sings in the autumn months,[2] and it is one of our earliest songsters, for, before the snowdrop appears, its loud and clear notes, which somewhat resemble those of the Blackbird, may be occasionally heard coming from the top of some tall tree,[3] even when the weather is stormy,

And fierce Aquarius stains th' inverted year.
THOMSON, *Winter.*

Hence one of the popular names of the bird is "The Stormcock."

During autumn and winter, when the Missel Thrush frequents grass fields and ground from which turnips have been removed, it is very wary; and immediately flies off when any attempt is made to approach it, uttering its harsh call-notes when on the wing, and occasionally after it has alighted on the top of some lofty tree in the neighbourhood. It is rather shy at all times, with the exception of the breeding season—after its eggs are hatched—when it becomes very bold, and attacks all birds and small animals which go near its nest, uttering loud screams,[4] and dashing at them

---

[1] "The Migration of Birds," a paper read to the Stirling Natural History Society on 24th March 1885, p. 7.

[2] Mr. Hardy writes, under date 24th November 1863:—"Missel Thrush in song on sea-banks." And again, on 27th December 1863:—"Missel Thrush singing in afternoon at Oldcambus."—*MS. Notes.*

[3] Dr. Stuart, Chirnside, records:—"The Missel Thrush in full song on the tree tops at the Manse," on 10th January 1884; and ten days afterwards:—"The Missel Thrush continues in full song, especially in windy, damp mornings, on the top of some tall ash tree."—*Hist. Ber. Nat. Club,* vol. x. p. 575.

[4] On this account it is called the "Screaming Mavis" in East-Lothian.

to frighten them away. There was a nest in a tall oak-tree at Paxton in the spring of 1872, and one morning my attention was attracted by the old birds screaming very loudly, flying round the tree, and dashing into it at intervals. On going forward I observed a squirrel near the nest, and the enraged parents kept darting down at it until they forced it to descend to the ground.

The Missel Thrush breeds early in the season, beginning to build about the end of March, or towards the middle of April, and choosing for the site of its nest the fork of an oak, elm, beech, or fir tree, in a wood or park, and often at a considerable height from the ground. It, however, sometimes selects a bush for its home; and as an example of this, I may mention that a pair built in one not above five feet high, on the lawn in front of my house, in the spring of 1874. The nest, which is large, is often placed in a conspicuous and exposed part of the tree, so that it is easily observed. It is composed of a variety of substances, such as moss, dry grass, straw, and coarse stems of plants, coated inside with a layer of mud, and lined with fine dry grass.[1] The eggs, which are four or five in number, are rather larger than those of the Blackbird, and are greenish or reddish white, spotted with brown and lilac. Two broods are often reared in the season.

The food of this Thrush consists of worms, slugs, and snails, with fruit and berries of various kinds, including those of the service tree, hawthorn, holly, and mistletoe, from the last mentioned of which it is said to derive its common name.

Although this bird is resident with us all the year as a species, many doubtless migrate southwards in autumn,

---

[1] Mr. Hardy, writing in 1840, says :—"One built its nest in the copestone of a dyke this summer in Bowshiel Dean. The lower part of the nest was composed almost entirely of the tree lichen, *Evernia prunastri*."—*MS. Notes.*

and return in spring. It has been frequently observed on migration at the lighthouses on the coasts of England and Scotland.[1] The late Mr. Carr-Ellison of Dunston Hill says: "It is pretty certain that the young of our English Missel Thrushes migrate largely into France, as there is a great accession of the species there in October and November. Their annual arrival is hailed by the bird-catchers, and by the epicures, with especial interest. A French cook will send up a dish of Missel Thrushes and Redwings in an irresistible form to the best tables, each bird enveloped in some delicate jelly of pearly hue."[2]

They suffered greatly, as did also Fieldfares, Redwings, and Song Thrushes, during the very severe frosts and snow-storms which occurred in the winters of 1874-75, 1878-79, 1879-80, and 1880-81. Mr. Hardy, Oldcambus, writing under date December 11th, 1874, says: "The great snow-storm had begun, but there was less snow here than in most places. *December* 19*th*.—Many Fieldfares, Redwings, and Missel Thrushes continued, for several days after this, to frequent the sea-banks at Siccar, where these were exposed to the melting influence of the sun's forenoon rays; working with all their might in digging at the softened clay, and in turning up old sheep-dung to procure insect food. *December* 28*th*.—Starlings fewer, as well as the Thrush kind. Some may have shifted place, but several died of hunger and cold; and of such I noted among the fallen, although they were not numerous, Missel Thrushes, Song Thrushes, and Fieldfares."[3]

Referring to my own notes, I find that, during the long continuance of the terribly severe weather in December

---

[1] *Hist. Ber. Nat. Club,* vol. vii. pp. 282, 283.
[2] See *Reports on the Migration of Birds,* 1879-86.
[3] *Hist. Ber. Nat. Club,* vol. v. p. 442.

1878 and January 1879,[1] a few birds of this species, which is usually very wild and shy, came, along with Song Thrushes, Redwings, Fieldfares, Blackbirds, House Sparrows, Hedge Sparrows, Redbreasts, and Titmice of several kinds, to a window-sill at Paxton House, and fed eagerly on crumbs of bread and meat which were put out for them every morning. The Missel Thrushes, Redwings, and Fieldfares were so tamed by the severity of the weather that they fed while people stood inside the window looking at them.

For some years after the severe winters of 1878-81 it seemed to be rather scarce in the county, but it has now apparently recovered its usual numbers. Darwin, in his *Origin of Species*, says that, where it becomes plentiful, it drives away the Song Thrush, being a stronger bird. This seems to have occurred in the neighbourhood of Chirnside about 1868, for we find Dr. Stuart writing in that year, that the Missel Thrush "is now so common that it has driven away the Common Thrush or Mavis altogether, or nearly so."[2]

---

[1] The winter of 1878-79 will be long remembered in the neighbourhood of Paxton as one of the most severe and protracted which has been experienced during the present century; for the oldest people in the village say that they do not recollect of a winter of such extraordinary severity, or of the snow lying so long upon the ground without a thaw. For nearly nine successive weeks in December, January, and February, snow lay deep all over the ground, and ponds and brooks were frozen. During that time also the Tweed was three times frozen over from bank to bank opposite Paxton House, so that men could walk across the river on the ice. The keenest frost occurred on the night of the 13th and early on the morning of the 14th December, when the thermometer, at the height of four feet from the ground, and fully exposed to the open air, marked no less than 40° of frost, or 8° below zero.

[2] *Hist. Ber. Nat. Club*, vol. ix. p. 332.

THE REDWING.

WIND THRUSH, SWINEPIPE.

Turdus iliacus.

See *Winter comes, to rule the varied year,*
Sullen and sad, with all his rising train;
Vapours, and clouds, and storms.
THOMSON, *Seasons.*

Byrdis flokkis ouer the fludis gray,
Unto the land sekand the nerrest way,
Quhen the cauld sessoun thame cachis ouer the see,
Into sum benar[1] *realme and warm countré.*
GAWIN DOUGLAS, *Virgil.*

THE Redwing is a regular autumn and winter visitor to this county, and generally makes its appearance from the northern parts of Europe, in our meadows and grass parks, about the end of October, where it may be seen feeding in small flocks with the Fieldfare. It sometimes comes earlier, however, and, in the autumn which preceded the severe winter of 1879-80, it was observed on migration at the Farne Islands as early as the 11th of September.[2] It remains with us the whole winter if the season be mild, and takes its departure in March and April for the countries near the Arctic Circle, where it breeds.[3]

---

[1] *Bene*—warm, genial.—Jamieson's *Scottish Dictionary.*

[2] Mr. Cordeaux records : "At the Longstone on 11th Sep., four Redwings killed at 3 a.m.; gloomy and thick."—*Report on Migration of Birds*: Autumn, 1879.

[3] Mr. Seebohm says :—"Though the Redwing does not usually nest in colonies like the Fieldfare, still it seems to prefer the society of its larger and more powerful relation, for wherever a colony of Fieldfares establish themselves, there, almost as surely, a pair of Redwings will build their nest close to them. In districts where trees abound, the Redwing seems to show a preference for the small

When it first arrives, and as long as the weather is open, it is wild and shy, but when keen frost sets in, with deep snow, both

> The Fieldfare grey, and he of ruddive wing,
> Hop o'er the fields unheeding, easy prey
> To him whose heart has adamant enough
> To level thunder at their humblest race.
>
> HURDIS.

The occurrence of a heavy snow-storm generally causes most of our Redwings to migrate further south. Those which from weakness or some other cause remain behind, maintain a struggling existence for a short time, when they may be seen frequenting the sheepfolds to pick up any little food which they can find there; but if the storm continue for some time, they very soon perish in the snow. "The severe snow-storms of 25th to 30th Jan., and 18th to 24th Feb. 1865," says Mr. Hardy, "accompanied as they were by vigorous frosts, made cruel havoc among the flocks of Redwings and Fieldfares that annually visit the Berwickshire coast. In the day-time, with feathers ruffled and taking short heavy flights, they frequented the sheepfolds to pick up scraps of turnip; but these afforded little assistance, and, as the storm increased, many crept into furze bushes and perished. After the snow had disappeared, the sides of sheltered deans—where, perhaps, they had roosted at night—were strewed with the dead bodies of the poor famished wanderers. About Oldcambus there were Fieldfares only, but from St. Abb's Head to Eyemouth, Redwings were intermingled in smaller numbers with their bulkier congeners. The mortality on that part of the coast had been

---

firs, where it builds its nest at no great altitude, and close to the stem. It is neatly made, and somewhat resembles that of the Ring Ouzel, though it is smaller and perhaps more firmly put together. The eggs are four to six in number, and the usual colour is pale bluish-green, thickly marbled over the entire surface with greenish brown."—Seebohm, *British Birds*, vol. i. p. 226.

very great. Each little nook of shelter in the deep ravines that communicate with the sea-side bore witness to it; the retired baylets and crannies of the rocks were scattered over with decayed feathered skeletons and along the Coldingham Sands their remains were frequently to be seen."[1]

The terrible severity of the winters of 1878-79, 1879-80, and 1880-81, of which some account has been given in the article on the Missel Thrush, destroyed great numbers of Redwings in Berwickshire; their remains being found in sheltered places, such as by the sides of hedges and dykes, and in deans, after the snow had disappeared.

This bird appears to be more tender than the Fieldfare or Missel Thrush, as it is generally the first which suffers when severe weather occurs, and

> The cherish'd fields
> Put on their winter-robe of purest white.
> THOMSON, *Winter*.

Its call-note is rather harsh, but its song is described by Linnæus in the account of his tour in Lapland as "delightful," and Mr. Hewitson says: "In our long rambles through the boundless forest scenery of Norway, or during our visits to some of its thousand isles, whether by night or by day, the loud, wild, and most delicious song of the Redwing seldom failed to cheer us."[2]

The food of this Thrush consists chiefly of worms, snails, and insects, and it also eats berries of various kinds. I have seen it greedily devouring rowans when they were ripe. When disturbed from its feeding-ground, the Redwing, like the Fieldfare and Missel Thrush, usually flies to the topmost branches of the nearest trees, the high elms above the Primrose Bank being one of its favourite resorts in the neighbourhood of Paxton, during the autumn months.

---

[1] *Hist. Ber. Nat. Club*, vol. v. pp. 232, 233.
[2] Hewitson, *Eggs of British Birds*, vol. i. p. 87.

THE REDWING.

The Redwing is very like the Song Thrush in its general appearance, but it may be easily distinguished by the reddish colour of the feathers under the wing, from which it derives its name, as well as by the white streak over the eye.

THE FIELDFARE.

FELTYFARE, FELDYFAR, FELTYFLEER, GREY THRUSH.

Turdus pilaris.

The Feltie, The Feltyfleer.

> *Where are your haunts, ye helpless birds of song,*
> *When winter's cloudy wing begins to shade*
> *The emptied fields; when ripening sloes assume*
> *Their deepest jet, and wild plums purple hang*
> *Tempting, yet harsh till mellowed by the frost?*
> *Ah, now ye sit crowding upon the thorns,*
> *Beside your former homes, all desolate,*
> *And filled with withered leaves; while Fieldfare flocks*
> *From distant lands alight, and, chirping, fly*
> *From hedge to hedge, fearful of man's approach.*
>   GRAHAME, *Birds of Scotland.*

THE first flocks of the Fieldfare generally arrive in the Merse about the end of October or beginning of November, and their appearance in our pasture fields is one of the many signs which we then have of the approach of winter, with its frosts and snows. Chaucer, in his *Assemble of Foules,* associates this bird with the characteristic weather of the winter season, for he calls it the " Frostie Feldefare," and Grahame, in his *British Georgics,* says—

> If, 'mid the tassels of the leafless ash,
> A Fieldfare flock alight, for early frosts prepare.

It comes to us chiefly from Norway and other northern countries of Europe, where it spends the summer, and nests on the pine and birch trees in colonies.

After its arrival here, and as long as the weather is mild, it may be seen feeding in considerable flocks on our pasture lands in company with Redwings, and occasionally a sprinkling of Missel Thrushes. It is then very wary, and a flock on being alarmed at once flies off to the top of the nearest high trees, or wings its way to some other favourite haunt at a distance.

When winter has fairly set in, and the first fall of snow occurs, its peculiar cry of "Yack-chuck-chuck-chuck,"[1] may be heard coming from some flock passing overhead; for it is then more frequently on the wing than usual, and its chuckling note readily catches the ear when—

> Thro' the hush'd air the whitening shower descends.
> THOMSON, *Winter*.

The Fieldfare feeds upon worms, snails, beetles, and insects of various kinds, as well as seeds of grasses and other plants, which it picks up in our pastures and stubbles as long as the weather is favourable. All kinds of berries form a favourite food in autumn and winter, and more especially those of the rowan tree[2] and the hawthorn. Its partiality for haws had been noticed by the poet Cowper for he says in the *Needless Alarm*—

> Not yet the hawthorn bore her berries red,
> With which the Fieldfare, wintry guest, is fed.

They suffered greatly from the effects of the unprecedented severity of the winters of 1878-79, 1879-80, and

---

[1] The popular name of the Fieldfare in the Orléanais is the "Chacha."—Rolland, *Faune Populaire de la France*, tome ii.; *Les Oiseaux Sauvages*, p. 237.

[2] The Mountain Ash (*Pyrus aucuparia*). The rowans remained long on the trees in the Lammermuirs in the Autumn of 1887, for, on December 2nd, while shooting with a friend on Cockburn Law, I saw a flock of Fieldfares feeding on the rowans of a leafless tree on the hill-side opposite The Retreat. Dr. George Johnston, writing of this tree, says:—"The Thrush tribe greedily devour the berries. I have never seen these used in Berwickshire as a bait to snare the birds, as Sir Robert Sibbald [*Scot. Illus.* ii. lib. 3, p. 6.] tells us was once the custom in Scotland: 'Ex setis caudæ (equorum) finguntur laquei, quibus appensis baccis Sorbi aucupariæ autumno Turdelæ, meruli et rubeculæ capiuntur.' In our moor districts the berries are called Reddens."—*Natural History of the Eastern Borders*, p. 80. For an account of the superstitions connected with the rowan tree, see *The Borderer's Table Book*, vol. vii. p. 183.

1880-81,[1] when the thermometer in Berwickshire fell as low as from 8° to 22° below zero. Many of them perished from hunger and cold, their remains, consisting of little else than bones and feathers, being found in sheltered spots, after the snow had disappeared. Mr. Hardy, writing of the effects of the winter of 1878-79, says:—"*December 9th*, 1878.—Snowstorm commenced, and continued more or less on the 10th, when, in the evening, great assemblages of Fieldfares arrived from the surrounding vicinage, to roost among the furze in the lower part of Oldcambus Dean. I was not abroad every day, but on the 14th the snow lay deep on the ground, and extended in one continuous sheet to the sea-side, except where washed by the tide. Fieldfares, in want of insects and worms, were greatly distressed, and from their dishevelled feathers appeared to be greatly pinched with the frosty air. Many of them kept in the folds all the day, hollowing out with their bills turnips which had been broken by the sheep. *December 16th.*—More dead Fieldfares. *December 17th.*—Frost still severe; Fieldfares were following Wood Pigeons in turnip fields, to profit by the morsels they left while picking holes in the turnips. They were in great extremity, hopping before me and tumbling over, with low dragging wings. Passing a wood-side, their mutilated remains, as well as those of the Redwing, were strewn wherever a sunny bank had tempted them with an offer of support."[2] It generally leaves for its breeding quarters in the north of Europe towards the end of April or beginning of May, but a few birds sometimes linger later, one being seen by me at Paxton in 1887 as late as the 15th of May. Mr. Seebohm says:—"The first visit to the

---

[1] On the night of the 13th Dec. 1878, the thermometer at Paxton fell to 8° below zero; and on 17th Jan. 1881, it fell at Blackadder to 22° below zero. In both cases the instruments were fully exposed to the weather.

[2] *Hist. Ber. Nat. Club*, vol. ix. p. 124.

breeding place of the Fieldfare is an event in the life of the ornithologist never to be forgotten. As you drive along the excellent Norwegian roads in the carioles or light gigs of the country, through the pine-forests or by the side of the cultivated land near the villages, there is little in the bird-life to remind you that you are not in one of the mountainous districts of England. As you approach the Dovrefjeld, however, the ground rises, the pines become smaller, and the hill-sides are sprinkled over with birch-trees. Now is the time to look out for the Fieldfare. Presently the long watched-for 'tsak-tsak' is heard. You tie your horse to the nearest tree, climb the hill-side whence the sound came, and presently you find yourself in a colony of Fieldfares. The birds make a great uproar as you invade their domain, but soon escape beyond gun-shot, and their distant 'tsak-tsak' is the only sound you can hear. Your natural impulse is to ascend the first tree where you can see a nest, which is almost sure to be placed in a fork of a birch-tree against the trunk and the first large branch. Close by are sure to be many more nests, some built in the flat horizontal branch of a pine; and outlying nests belonging to the colony will be found for some distance all round. The nest is very similar to the Blackbird's or the Ring Ouzel's in construction and materials. The eggs are from four to six in number, the average type resembling a very handsome Blackbird's egg."[1]

The subject of our notice has been celebrated as a luxury for the table, and is supposed to have been the Thrush so highly esteemed by the Romans, which was fattened by a paste of figs and flour.[2] It is often cooked for the table when it is shot in autumn, and is excellent eating at that season.

---

[1] *History of British Birds*, vol. i. 1883, pp. 229, 232.
[2] See Macgillivray's *British Birds*, vol. ii. p. 111.

WHITE'S THRUSH.

Turdus varius.

Ay farest faderis hes farrest fowlis,
Suppois thay haif no sang bot youlis.

W. Dunbar.

The only specimen of this very rare bird which has been obtained in Scotland, was got at Hardacres, in the Parish of Eccles, Berwickshire, as recorded by Mr. Andrew Brotherston, who, writing in the *History of the Berwickshire Naturalists' Club*,[1] says:—"During the last week of December 1878, a specimen of this very rare and beautiful thrush was shot by Mr. Forbes Burn at Hardacres, Berwickshire. Not being aware of its rarity, unfortunately only a portion of the bird was saved,—the head and wings unskinned, with part of the skin of the breast and back,—and forwarded to me on the 22nd January following, to preserve as an ornament for a lady's hat. I immediately took the necessary steps to try and secure what was left of it for the Ornithological Collection of the Tweedside Physical and Antiquarian Society, which were successful, the owner very kindly and promptly presenting it to the Society. Mr. Burn told me that it resembled a hawk when on the wing; and that some small birds which were feeding on the ground took flight at its approach. I am much indebted to Professor Newton, who has seen and examined this specimen, for a large amount of interesting information

---

[1] *Hist. Ber. Nat. Club.* vol. viii. pp. 518, 519, 520.

concerning White's Thrush. He writes :—' At least nine examples of this bird have been before now killed in Britain. They are :—

    1. Christchurch, Hants, 24th Jany. 1828.
    2. Bandon, Cork, Dec. 1842.
    3. Welford, Warwickshire, 26th Jany. 1859.
    4. Ballymahon, Longford, 1867.
    5. Hestercombe, Somerset, Jan. 1870.
    6. Langsford, Somerset, 6th Jan. 1871.
    7. Hickling, Norfolk, 10th Oct. 1871.
    8. Castle Eden, Durham, 31st Jan. 1872.
    9. Probis, Cornwall, early in Jan. 1874.

On comparison with a specimen that has been long mounted, the fresh beauty of the colours in yours is very decidedly marked, yet I fear nothing can be done to preserve its tints, and that, when as many years have elapsed, their richness will have disappeared. I have wholly failed to find any indication that would enable me to determine the age or sex of your bird. There is no question about it being the true *T. varius* of Pallas, though the tail is wanting,—an unfortunate thing, as therein lies one of the most curious characters of this species, one that is possessed, so far as I know, by only one other species of Thrush (*T. Horsfieldi*)—the presence of fourteen instead of twelve rectrices. The real White's Thrush (*T. varius, Pallas*), was first described as an inhabitant of Siberia, to which country and to the north-east of Asia (*i.e.* China and Japan) it is now known to be a regular summer visitant. Owing to causes that I cannot attempt to explain, a small number of examples seem yearly to migrate westwards in autumn, and to come into Europe, where they occur as stragglers; but the majority, no doubt, retire more or less due southwards, for they have been obtained in winter in the Philippine Islands and such like places.'"

Mr. Seebohm says :—" The occurrence of White's Thrush

in Europe can only be considered accidental, though accidents of this kind happen regularly. After the breeding season is over in the Arctic regions, the great stream of migration which passes from north to south through Central Siberia appears to divide before it reaches the mountains of Mongolia, to avoid the deserts beyond. Some species of the birds turn east and others west; and of the species which Nature has ordained to winter east, some individuals, probably for the most part young birds who have never migrated before, lose their way and get into the wrong stream, and thus find their way into Europe as strangers from the East, some of whom fall into Gaetke's hands on Heligoland every year. The breeding ground and true home of this fine Thrush is South-Central and South-Eastern Siberia and North China. It winters in South Japan, South-West China, and the Philippine Islands, occasionally straying as far west as Sumatra."[1]

This bird was named White's Thrush in honour of the Rev. Gilbert White, whose memory is dear to ornithologists, and whose *Natural History of Selborne* is a perennial source of delight to all lovers of Nature.

---

[1] Seebohm, *British Birds*, vol. i. 1883, pp. 200, 201.

THE BLACKBIRD.

OUZEL, GARDEN OUZEL, MERLE.

Turdus merula.

The Blackie.

> Oh! when my friend and I
> In some thick wood have wandered heedless on,
> Hid from the vulgar eye, and sat us down
> Upon the sloping cowslip-covered bank,
> Where the pure limpid stream has slid along
> In grateful errors through the underwood,
> Sweet murmuring; methought! the shrill-tongued Thrush
> Mended his song of love; the sooty Blackbird
> Mellowed his pipe, and softened every note.
> BLAIR, *The Grave.*

> The Merle, in his noontide bow'r,
> Makes woodland echoes ring.
> BURNS, *Queen of Scots.*

THE rich flute-like notes of the Blackbird " telling his love-tale to his mate," in the spring and early summer months, although they lack the variety of those of the Mavis,[1] give an additional charm to the music of our woods and dells, and render it one of our most favourite birds.

Its loud, clear, and melodious song is heard with the best effect in some of our wooded deans, on a calm, sunny

---

[1] This seems to have been observed as early as the time of Sir David Lindsay of The Mount, who is supposed to have written *The Complaynt of Scotland* about 1548, for he says in it :—
> The Maveis' maid myrtht for to mok the Merle.

morning or evening, in the month of May,[1] for, on such occasions—

> the Merle's note,
> Mellifluous, rich, deep-toned, fills all the vale,
> And charms the ravished ear.—GRAHAME.

The Blackbird is very frequently mentioned by our poets; one of the earliest poems of Dunbar, who may be styled the Burns of the fifteenth century, being "The Merle and the Nichtingale," beginning—

> In May, as that Aurora up did spring,
> With crystal ene chasing the cluds sable,
> I hard a Merle, with merry notes sing
> A sang of love.

Coming to a much later date, we find Sir Walter Scott referring to its song, in the opening verses of the "Gathering," in *The Lady of the Lake*—

> The summer dawn's reflected hue
> To purple changed Loch Katrine blue;
> Mildly and soft the western breeze
> Just kiss'd the lake, just stirr'd the trees :
> . . . .
> The Blackbird and the speckled Thrush
> Good morrow gave from brake and bush.

And Leyden, describing the sylvan beauties of the Borderland, says—

> The mountain ash whose berries shine :
> The flaxen birch that yields the palmy wine ;
> The guine, whose luscious sable cherries spring
> To lure the Blackbird 'mid her boughs to sing.

---

[1] Gower, in a Sonnet on the month of May, written in the reign of Edward III., mentions the singing of the Merle as one of the chief attractions of this charming season of the year :—

> Pour comparer le joli mois de Mai,
> Je (le) dirai semblable à Paradis ;
> Car lors chantoit et Merle et Popegai ;
> Les champs sont verds, les herbes sont fleuries,
> Lors et Nature dame du pais.
>
> *Specimens of Early English Poetry*, vol. i. p. 170.

The Scottish poet, Alexander Scott, who wrote in the time of Queen Mary, says in *The Blait Luvar* :—

> Quhen flora had ourfret the firth,
> In May of every moneth quene,
> Quhen Merle and Mavis sings with mirth
> Sweet melling in the schawis schene.

The Blackbird is found in Berwickshire wherever there are trees, but in the upland districts it is by no means so plentiful as in the well-wooded parts of the Merse. Its favourite resort is the neighbourhood of gardens and shrubberies which are bordered by lawns and grass parks, and it also loves to frequent our bosky deans, through many of which small streams of water meander. Numbers also frequent tall bushy hedgerows in open weather during autumn and winter. It often roosts at night during the latter season in young fir woods.[1] It is evidently a partial migrant in this county, for many more are seen here during spring, summer, and autumn, than in the later months; but it never entirely leaves us, for a few remain here even through the severest winters. In autumn and spring it is frequently seen while on migration, at the Lighthouses on the Farne Islands and on the Isle of May,[2] and it is probable that, as in the case of the Song Thrush, great numbers of our homebred Blackbirds, after being joined in autumn by those which have come from the north of Europe to our district, leave for the south on the approach of winter, and return again in spring. Local migration also appears to take place, numbers of this species betaking themselves in hard weather to milder parts of the country.[3]

---

[1] When shooting Wood Pigeons with Mr. Pringle, Ayton Castle, in the young fir woods there, on the evening of 30th Dec. 1887, I observed great numbers of Blackbirds coming in to roost in the trees about 4.25 p.m. They were constantly flying about the trees and "pinking."

[2] The following are the dates upon which Blackbirds were seen on migration at the Farne Islands and Isle of May :—*Autumn* 1879.—Farnes, Oct. 15th to Dec. 5th. *Autumn* 1880.—Farnes, Oct. 21st to Nov. 26th. *Spring* 1881.—Farnes, May 22nd. *Autumn* 1881.—Farnes, Oct. 2nd; Isle of May, Oct. 14th to 24th. *Autumn* 1882.—Farnes, and occurred throughout season. *Spring* 1883.—Isle of May, March 2nd. *Autumn* 1883.—Isle of May, Oct. 13th to 15th, and throughout season. *Spring* 1884.—Isle of May, Feb. 20th to April 30th ; Farnes, March 14th to 20th. *Autumn* 1884.—Isle of May, Sep. 11th ; Farnes, October and November to 19th Jan. 1885.—Extracts from *Reports on the Migration of Birds*, from 1879 to 1884.

[3] See "Effects of recent Winter Storms," by Mr. Robert Gray.—*Hist. Ber. Nat. Club*, vol. ix. p. 499.

The Blackbird is very hardy, and does not readily succumb even in the severest weather, when, accompanied by long-continued frost, a "lying-storm" sets in, and the—

> Blinding flakes, by day, by night,
> In thickening showers descend.
> GRAHAME, *January.*

In the terrible winters of 1878-79, 1879-80, and 1880-81, however, numbers perished, while they were noticed to be scarce in the following springs,[1] and did not recover their ordinary numbers until about 1883. Piebald and albino examples are occasionally seen in the county, but are of rare occurrence.[2] Major Dorling, Duns, informs me that a cream-coloured bird, with a yellow bill, frequented the shrubbery at the south end of Duns Castle lake in the spring of 1886; and Mr. Charles Watson mentioned to me that he had sometimes noticed the same bird in his garden. It was caught under a sieve or "riddle" by a person in Duns, in the end of December 1886, and given to Mr. Hay of Duns Castle, who kept it for some months in an aviary, where it died. Mr. John Fulton, salmon fisher, Milne Graden, told me on the 17th of January 1888, that, during the fishing season of 1887, a white or cream-coloured Blackbird frequented "Keppie" Island, which is below "Dreeper" Island, on the Tweed at Milne Graden. An ash-coloured specimen is recorded as having been seen by Dr. Stuart of

---

[1] Mr. Hardy, Oldcambus, in a paper on the effects of the winter of 1878-79, mentions that several Blackbirds were found dead in his locality.—*Hist. Ber. Nat. Club.* vol. ix. pp. 123-130. On referring to my own notes on the subject, I observe that one or two were found dead in the neighbourhood of Paxton, in January 1881, from the effects of the severe weather. Mr. Hardy, writing on 8th Feb. 1879, says:—"Three Blackbirds appeared in the garden—the only ones left," and adds, that in the spring of 1880 they had not recovered their old numbers. Dr. Stuart, Chirnside, writes under date 6th March 1879:—"The winter has been unprecedentedly severe and protracted: Blackbirds and Thrushes have got a great thinning."—*Hist. Ber. Nat. Club.* vol. ix. pp. 133-136.

[2] A popular remark in France when any one promises to undertake an impossible task is: "Si vous faites cela je vous donnerai un Merle blanc."—Rolland, *Faune Populaire de la France,* t. ii.; *Les Oiseaux Sauvages,* p. 248.

Chirnside, near the Pistol Plantings on the 5th of March 1883;[1] and a pied bird was got at Lochton, in the parish of Eccles, in 1872.

The food of this species consists chiefly of earth-worms, snails, and grubs of various kinds, with the addition of garden fruit in summer and autumn, and wild berries, such as those of the ivy,[2] rowan, and elder. It is very fond of strawberries and cherries. On this account it is much disliked by gardeners, although it undoubtedly makes amends for at least a part of its depredations, by the destruction of great numbers of noxious grubs and slugs.

The Blackbird breaks the shells of snails, like the Song Thrush, by seizing them in its bill and striking them against a stone. In autumn it is often seen in considerable numbers amongst turnips, where it feeds on worms and insects in the drills under the leaves, and whence, on being disturbed, it generally flies rapidly away to the nearest hedge, uttering its cry of alarm. It also frequents bean fields. In winter haws form a favourite food, and in severe weather it frequents the neighbourhood of houses and stackyards, where it picks up waste grain and seeds of various kinds. When snow is lying deep on the ground, and other small wild birds are being fed with crumbs at our windows and are crowding around the morsels put out for them, I have sometimes observed that

> At a distance on the leafless tree,
> All woe-begone, the lonely Blackbird sits:
> . . . . .
> Full oft he looks, but dare not make approach,[3]

until, being satisfied that there is no danger, he ventures forward for his share of the feast.

The Blackbird does not congregate in flocks under ordi-

---

[1] *Hist. Ber. Nat. Club*, vol. x. p. 573.
[2] Writing under date May 26th, 1866, Mr. Hardy says:—"Ivy berries ripe in Pease Dean and Lumsden Dean—Blackbirds feasting on them."—*MS. Notes.*
[3] J. Baillie, "A Winter's Day."

nary circumstances, but is often seen in company with those of other birds, such as Starlings, Redwings, Fieldfares, Larks, and Plovers, while on migration.[1] It does not begin to sing in spring so early as the Thrush, but its notes are occasionally heard in January. It is often one of the first songsters of the morning.[2] On the 15th of May 1887, it began its song in my garden as early as 3 A.M., and was immediately followed by a Thrush, and a chorus of other birds. It is sometimes kept as a cage-bird, and can be taught to whistle airs. In my boyhood a Blackbird reared by Mr. John Tait, tailor, in the village of Salton, East-Lothian, used to delight the villagers by whistling "Ower the water to Charlie," and one which I had at Paxton for some years, repeated the same tune with wonderful precision. As a pet it has the disadvantage of requiring its cage to be constantly cleaned. Mr. Hardy likens some parts of the natural song of this bird to the words—"Cock your periwig," "Tie your cravat."

It builds its first nest very early in spring,[3] generally selecting a thick bush, such as a yew or young spruce fir tree, for the purpose. It is also partial to old ivy on walls, and later in the season, when the leaves are out, it often selects a well-concealed spot in a hawthorn bush or hedge for rearing its second brood.

The nest is usually constructed of coarse grass and twigs, with some moss and dry leaves, cemented together with mud, and the inside lined with fine dry grass. The eggs, which are four or five in number, are generally of a bluish green, thickly speckled with brown.

---

[1] See *Reports on the Migration of Birds*, 1879-84.

[2] Dr. Stuart heard the Blackbird's spring notes at Chirnside on 19th Jan. 1882. When returning from Billie Mains on 6th July 1883, he heard it begin to sing at 2.15 A.M.—*Hist. Ber. Nat. Club*, vol. x. pp. 572, 574.

[3] A Blackbird's nest, with one egg, was found near Duns in the end of January 1888.—*Berwickshire News*, 31st Jan. 1888.

Like the Song Thrush, the Blackbird is a great favourite with boys on account of its early nesting in spring, and they love to search for the nest and handle the speckled eggs. It is to be hoped that but seldom in Berwickshire

> The truant schoolboy's eager bleeding hands,
> Their house, their all, tears from the bending bush.
> GRAHAME, *Birds of Scotland.*

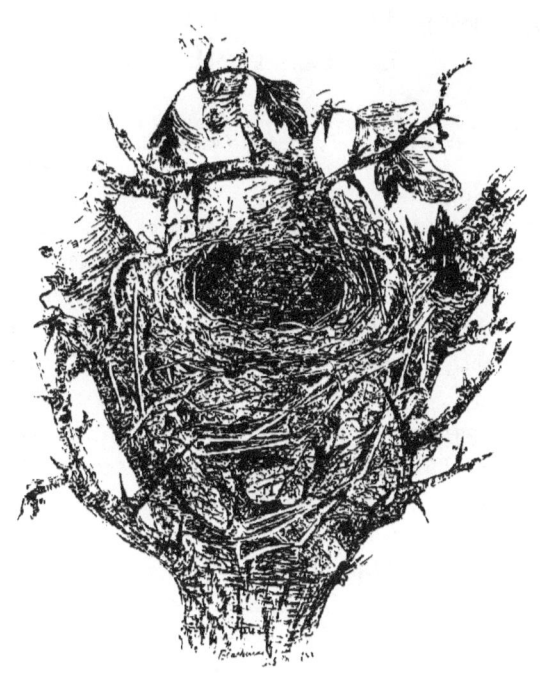

THE RING OUZEL.

ROCK OUZEL, MOUNTAIN OUZEL, TOR OUZEL, WHITE-BREASTED BLACKBIRD, MOOR BLACKBIRD.

Turdus torquatus.

The Ring Blackbird, The Moor Blackbird, The Rock Starling.[1]

The soote season, that bird and bloom forth brings,
With green hath clad the hill and eke the vale.
LORD SURREY.

WITH the return of spring the Ring Ouzel makes its appearance in small numbers in the Lammermuirs, where it usually arrives in the month of April;[2] and, after spending the summer there, it migrates towards the south in September and October, although one or two occasionally remain a little later.

In appearance it is very like a Blackbird, with a white crescent on the breast, but it is a much wilder bird than our yellow-billed favourite, being very shy and wary, except when near its nest in the breeding season.

It is found frequenting rocky deans and cleughs in remote districts of the Lammermuirs, and seems to be partial

---

[1] It is called the Rock Starling about Byrecleuch.
[2] Mr. Hardy records its arrival near Oldcambus, on 24th April 1872.—*Hist. Ber. Nat. Club*, vol. vi. p. 383. 7th April 1874—On heights above Dowlaw.—*Ibid.* vol. vii. p. 279. 21st April 1876—In the Dean near Oldcambus.—*Ibid.* vol. viii. p. 153. 15th April 1879—In Oldcambus Dean.—*Ibid.* vol. ix. p. 129. 18th April 1883—On Bell Hill.—*Ibid.* vol. x. p. 564. 25th April 1884—Dowlaw Dean.—*Ibid.* vol. x. p. 570.

to the neighbourhood of the upper streams of the Dye and the Watch, where its song, which somewhat resembles that of the Blackbird, adds a charm to the moorland scenery.[1] Mr. James Smith, shepherd, Byrecleugh, has informed me that the Ring Ouzel is often to be seen in the Hall Burn, Wood Cleugh, and near Wrink Law, and that he has several times discovered its nest in that locality. Mr. Hardy writes:—"It used to come, when I was a boy, in flocks in spring-time, to the haugh beside the Pease Burn below Penmanshiel, where a large quantity of haws on an old hedge formed the attraction. It bred abundantly on Ewieside about 1856, and frequented Channobank, where there were whin bushes, on the tops of which it used to be seen whistling. I have observed it on the Weir Burn, on the east side of Laughing Law. It breeds in many wooded glens of the Lammermuirs. On 22nd July 1885 I observed several on the 'Glitters' on Great Dirrington Law. Dowlaw Dean,[2] near Fast Castle, appears to be a favourite haunt of the Ring Ouzel on its arrival in spring, as it has been often noticed there in April. Mr. John Boyd, of Cherrytrees, tells me that it used to be so plentiful about the moors at Abbey St. Bathans during the grouse-shooting season, that it frequently put the dogs off the scent of the game."[3] Mr. Lockie states that it is rather scarce in the Westruther and Spottiswoode districts, but is frequently seen between

---

[1] I love to see the purple bloom across the moorlands spread,
The purple bloom o' Lammermuir, I love, I love to tread;
I love to wander by the marsh that plough has never torn,
I love to see the reed-tufts wave where ne'er has waved the corn.
　　　　　　　　　Dr. HENDERSON, Chirnside, *MS. Poems.*

[2] No more in Dowlaw's rugged Dean we'll seek the rose-wort's flower,
Nor sit beside the murmuring brook at noon-tide's sultry hour.

From MS. Lines to the Memory of Alexander Allan Carr, Surgeon, Ayton, Author of the *History of Coldingham Priory*, by GEORGE HENDERSON, Surgeon, Chirnside, 10th March 1840.

[3] Mr. Hardy's *MS. Notes.*

Blyth and Broadshawrig,—also that it is rare in the neighbourhood of Earlston, although it is occasionally found on the Black Hill. In his account of the birds of Lauderdale, Mr. Kelly says that on Longcroft Water, far up among the junipers, Ring Ouzels have their stated visits every year, and that they are clamorous when they are approached in the breeding season.[1] Mr. J. L. Mack, Coveyheugh, informs me that, in the first week of September 1885, he noticed a dozen of these birds lying dead in the garden of the Rev. R. B. Smith, minister of Cranshaws, who had shot them while they were destroying his fruit. A specimen was observed near the "Cockit Hat" Plantation on Lamberton Moor, in April 1883, by the gamekeeper at Mordington. It does not often visit the lower parts of the county. Miss Georgina S. Milne-Home saw one on the lawn at Milne Graden, about the beginning of October 1885, which would probably be on migration southwards.

The food of this species consists of worms, snails, slugs, beetles, and other insects, with fruit and various wild berries, including those of the rowan tree (*Pyrus aucuparia*), blae-berry (*Vaccinium myrtillus*), and cranberry (*Vaccinium oxycoccus*). Colonel Brown of Longformacus has told me that it feeds in great numbers on the rowan berries at Dye Cottage in September and October, and that it leaves the moors as soon as they are done. It also eats haws, and helps itself to the fruit in gardens near its haunts.

The nest of the Ring Ouzel, which is somewhat like that of the Blackbird, is often placed on the ground, amongst bushes, in such situations as a bank by the side of a stream; but it is occasionally built under a rocky ledge, or in a stunted tree. The eggs, which are four or five in number, are bluish green, blotched with reddish brown, and are

---

[1] *Hist. Ber. Nat. Club*, vol. vii. p. 303.

generally more distinctly marked than those of the Blackbird, the spots being clearer and not so much run together. Mr. James Smail, in describing a nest which he found while fishing in Earnscleuch Glen, in the Lammermuirs, on the 28th of June 1869, says :—" Both parents were in a great state of chatter and excitement, and showed a great deal of boldness by flying at me repeatedly as I handled the young, which were just 'ripe.' The cock flew oftenest at me, but of course, both birds swerved when they came near my face. The nest was placed under the branches of a large bunch of heather growing on the top of a rock, which stands about ten feet above the burn in which I was fishing."[1]

---

[1] *Hist. Ber. Nat. Club*, vol. vi. p. 96.

THE WHEATEAR.

WHITE-RUMP, FALLOW SMICH, WHITE-TAIL, STONECHAT, STANECHACK.

Saxicola œnanthe.

𝕿𝖍𝖊 𝖂𝖍𝖎𝖙𝖊-𝕽𝖚𝖒𝖕, 𝕿𝖍𝖊 𝕾𝖙𝖆𝖓𝖊𝖈𝖍𝖆𝖈𝖐𝖊𝖗.[1]

> *A good Westphalia gammon*
> *Is counted dainty fare,*
> *But what is that to salmon*
> *Just taken from the Ware?*
> *Wheatears and quails,*
> *Cocks, snipes, and rayles,*
> *Are prized while season's lasting,*
> *But all must stoop to craw-fish soup,*
> *Or I've no skill in tasting.*
>
> W. H. LOGAN, *The Fisherman's Song in Pedlar's Pack of Ballads.*

MR. HARDY, in his interesting papers on the "Arrival, Departure, and Local Migration of Birds near Oldcambus,"[2] very often refers to the Wheatear, for it is one of the most conspicuous migrants near the coast, and in the higher districts of the county. It generally arrives in Berwickshire from the south in the end of March or beginning of April, and departs southwards again in August and September.[3]

---

[1] Called "Stanechacker" by the Coldingham fishermen.—J. Hardy in *Hist. Ber. Nat. Club*, vol. ix. p. 393.

[2] See *Hist. Ber. Nat. Club*, vols. vi., vii., viii., ix., and x.

[3] Wheatears were observed on migration in *Autumn* 1880.—At Farnes on 10th Aug. and 15th Sep. *Spring* 1881.—Farnes, 25th March and first week in Ma . *Autumn* 1881.—Isle of May, 22nd Aug. to 22nd Sep. *Spring* 1882.—Isle of May, 29th March; Farnes, 22nd March. *Autumn* 1882.—Farnes, 12th August.

Shortly after its arrival in spring, while on its way to the uplands, it may be occasionally seen on ploughed fields near the sea-coast, and in the inland and lower parts of the county, flitting from clod to clod, displaying the conspicuous white patch above the tail which has gained for it the name of "White-rump." Like the Stonechat and the Whinchat, it is also fond of flying along palings and stone-dykes and alighting on their most prominent points, where its lively movements and handsome appearance seldom fail to attract the notice of the most casual observer. During summer it frequents the wild pastures and open moors, where it breeds, making its nest, which is composed of fine grasses mixed with wool, moss, feathers, and hair, in a hole in some dry-stone or turf dyke, or in a rabbit burrow, and laying five or six pale blue eggs. I have frequently seen the Wheatear in summer on the moorland road from Wedderlie to Dye Cottage, and also on the moors along the course of the Dye, and on the road from Cattleshiel to Westruther. Mr. Hardy mentions that it breeds along the skirts of the moors by the sea-coast near Oldcambus;[1] and the Rev. George Cook, Longformacus, informs me that its nest, with eggs, has been found near Whitchester by Mr. Smith. Young Wheatears were observed at Oldcambus, and above Head-chesters, and on the post road above Upper Moorhouse, in July 1882.[2]

Shortly before its departure in autumn it is sometimes seen in the lower parts of the county, flitting about

---

Spring 1883.—Isle of May, 5th and 13th April; Farnes, 2nd to 24th April.
Autumn 1883.—Isle of May, 2nd and 3rd Sep. ; Farnes, 2d and 3rd Sep.
Spring 1882.—Isle of May, 26th March to 3rd April; Farnes, 19th March.
Autumn 1884.—Isle of May, 15th Sep. to 22nd Oct. ; Farnes, 6th Aug. to Oct.
Spring 1885.—Isle of May, 20th April; Farnes, 5th April. *Autumn* 1885.-Farnes, 7th and 9th Aug. *Spring* 1886.—Farnes, 22nd Feb.—From *Reports on the Migration of Birds*, 1879-85.

[1] *Hist. Ber. Nat. Club*, vol. ix. p. 392.
[2] *Ibid.* vol. x. p. 560.

dykes and palings, apparently preparing for its southern journey.

> When, in a thousand swarms, the summer o'er,
> The birds of passage quit our English shore,
> By various routes the feather'd myriad moves;
> The Beccafico seeks Italian groves,
> No more a Wheatear, while the soaring files
> Of seafowl gather round the Hebrid isles.
>
> CHARLOTTE SMITH.

The Wheatear is much esteemed as a delicacy for the table, and the following account of the method of taking it in traps is given by Yarrell:—" The well-known South Downs of Sussex are visited by the Wheatear from the end of July to the middle of September in vast numbers, consisting almost exclusively of the young birds which, having been bred in other parts of the country, or perhaps even further to the northward, are then pressing forward on their autumnal journey. Being at that time fat and of excellent flavour, these periodical emigrants are in great request as a delicacy among those who frequent the many watering-places on that coast. The birds are chiefly supplied by the shepherds, who set traps for them on the downs over which their flocks graze. The Wheatear trap is formed by cutting an oblong sod of turf from the surface, about eight inches by eleven, and six inches thick, which is taken up and laid in the contrary way, both as to surface and direction over the hole, thus forming a hollow chamber beneath. Besides this chamber, two other openings are also cut in the turf, about six inches wide and of greater length, which lead into the chamber at opposite ends, that the bird may run in under the turf through either of them. A small straight stick, sharpened at both ends, not very unlike a common match, but stouter, is fixed in an upright position a little on one side of the middle of the square chamber; the stick supports two open running loops

of twisted horse-hair placed vertically across the line of
passage from either entrance to the opposite outlet, and the
bird attempting to run through is almost certain to get its
head into one of these loops and be caught by the neck:
upon the least alarm, even the shadow of a passing cloud,
the birds run beneath the clod and are taken. However
inefficient these traps may appear to be from the description,
the success of the shepherds is very extraordinary. One
man and his lad can look after from five to seven hundred
of them. They are opened every year about St. James's
Day, July 25th, and are all in operation by August 1st.
The birds arrive by hundreds, though not in flocks, in daily
succession for the next six or seven weeks, probably de-
pending on the distance northward at which they have
been reared. The season for catching is concluded about
the end of the third week in September, after which very
few birds are observed to pass. Pennant, more than a
century since, stated that the numbers snared about East-
bourne amounted annually to about 1840 dozens, which were
usually sold for sixpence the dozen, and Markwick, in
1798, recorded his having been told that, in two August
days of 1792, his informant, a shepherd, had taken there
twenty-seven dozens; but this is a small number compared
with the almost incredible quantity sometimes taken, for
another person told the same naturalist of a shepherd who
once caught eighty-four dozens in one day. In Montague's
time (1802) the price had risen to a shilling the dozen, and
it is now much higher, through the greater demand for and
smaller supply of the birds."[1]

The food of this species consists of insects, worms, and
grubs of various kinds.

---

[1] Mr. Hardy writes that it is occasionally caught accidentally at the mouths
of rabbit-holes, on the coast near Oldcambus, in traps set for rabbits.—*MS.
Notes.*

The song of the male is pleasant, though not loud, and is often uttered when hovering on the wing near its nest.[1]

---

[1] Mr. Hardy mentions that he heard it singing on the tops of whin bushes in a field called the "Red Cot Waas," near St. Helen's Church, on 1st May 1874.—*MS. Notes.*

THE WHINCHAT.

THE FURZE-CHAT, WHIN OR FERN LINTIE, WHIN CLOCHARET.

Saxicola rubetra.

The Whinchacker.[1]

I have heard
Where melancholy Plovers hovering screamed,
The Partridge-call, at gloamin's lovely hour,
Far o'er the ridges break the tranquil hush ;
And morning Larks ascend with songs of joy,
Where erst the Whinchat chirped from stone to stone.[2]

GRAHAME, *British Georgics.*

THE Whinchat is a regular summer visitor to Berwickshire, generally arriving about the end of April or the beginning of May,[3] and departing towards the south in September and October.

It is found in much the same localities as the Stonechat, which it resembles in its habit of constantly flitting from the top of one whin bush, tall thistle, or other moorland weed to that of another, or from stone to stone, on the moors. It also inhabits the more cultivated districts, where it may be often seen by road-sides flying along dykes and

---

[1] Dr. Johnston's *MS. Notes.*

[2] The Poet here refers to the improvement of moorland by ploughing and cultivation, which, in the opinion of many experienced hill-farmers of the present day, has been carried out to far too great an extent.

[3] Mr. Hardy records its arrival in the neighbourhood of Oldcambus on 3rd May 1875 (*Hist. Ber. Nat. Club*, vol. vii. p. 485) ; 2nd May 1879, at Cockburnspath (*ibid.* vol. ix. p. 393) ; 7th May 1881 (*ibid.* p. 555) ; 25th April 1882 (*ibid.* vol. x. p. 558) ; 6th May 1883, on Earnslaw and Greenside Hills, and on Lowrie's Knowes near Dowlaw (*ibid.* p. 565).

wire fences, and alighting upon their highest points, uttering from time to time its call-note of " Utik-utik-tik-tik-tik."

When migrating southwards in autumn, it is occasionally observed on palings, hedges, or stooks, in the lower parts of the county, where it adds an interesting variety to the usual birds in the Merse at that season.

Mr. Hardy records the following localities as favourite haunts and breeding-places of the Whinchat :—The Seabanks east from Redheugh, Oldcambus Dean, Pease Bridge, Edmonsdean, Haugh near River Eye at Quixwood, head of Allerburn near Abbey St. Bathans, Windshiel, ground behind Cockburn Law, Whitchester and Rigfoot, Penmanshiel Moor, Soldier's Dyke near Dowlaw, Earnslaw and Greenside hills, Lowrie's Knowes near Dowlaw, and Redheugh Hill.[1] He mentions that the Whinchat and the Stonechat were once much more numerous than at present on the skirts of the moors in the neighbourhood of Oldcambus, but cultivation has driven them into the back wastes.[2]

Dr. Stuart found it in considerable numbers in Gordon Bog on the 30th of June 1879,[3] and it is also recorded as having been seen at Corsbie Bog on the same day.[4] It was observed near Longformacus on the occasion of the Meeting of the Berwickshire Naturalists' Club there on the 26th of July 1882.[5] On the 5th of May 1885, while on my way to Greenlaw, I noticed several birds of this species flitting along the top of the hedges by the sides of the public road near Crossrigg Farm, and I saw one or two perching on the wire fence by the side of the road leading from Burnhouses to Abbey St. Bathans in May 1886.

The song of the Whinchat, which may be heard shortly

---

[1] See *Hist. Ber. Nat. Club*, vols. vii., viii., ix., and x.
[2] *Ibid.* vol. ix. p. 393. [3] *Ibid.* vol. ix. p. 230.
[4] *Ibid.* vol. ix. p. 235. [5] *Ibid.* vol. x. p. 22.

after its arrival in spring, when the bird is perched on the top of a whin bush or fluttering in the air, is sweet and agreeable. Mr. Hardy records hearing a male singing very sweetly on Redheugh Hill on the 17th of May 1882;[1] and I heard several on the whinny ground by the side of the Whitadder, opposite Ellemford Smithy, on the 12th of May 1887. The partiality of this bird for whins is alluded to by Charlotte Smith in the following lines:—

> For 'midst the yellow bloom the assembled Chats
> Wave high the tremulous wing, and with shrill notes,
> But clear and pleasant, cheer the extensive heath.

Its food consists to a great extent of gnats and other insects, to seize which it may be often seen leaping into the air from its perch. It likewise feeds on small worms, beetles, and wireworms.

On the moors and uplands, the nest, which is generally built of moss and dry grass lined with rootlets and horsehair, is placed on the ground amongst heather, or under a whin bush; while in meadows a thick tuft of rank grass is often chosen for its site. The eggs, which are five or six in number, are greenish blue, and are usually marked at the larger end with fine specks of pale reddish brown.

The Whinchat can be distinguished when in mature plumage, from the Stonechat and Wheatear, which it somewhat resembles in general appearance, by its mottled brown head and back, and a white streak over the eye, combined with a white patch on each side of the neck.

---

[1] *Hist. Ber. Nat. Club*, vol. x. p. 559.

THE STONECHAT.

STONECHATTER, STONESMICH, BLACKY-TOP.

Saxicola rubicola.

𝕮𝖍𝖊 𝕾𝖙𝖆𝖓𝖊𝖈𝖍𝖆𝖈𝖐𝖊𝖗.[1]

The Stainyell and the Schackerstane,
Behind the lave wer left alane,
With waiting on thair marrows.
     BURREL'S *Passage of the Pilgrimer.*

THIS interesting little bird, which is generally a permanent resident in the county, is found on many of our rough, stony moors, particularly those which are interspersed with whins, junipers, and other similar bushes. During very severe weather in winter it seeks more sheltered spots near the sea-coast, such as Oldcambus Dean, and some may even migrate southwards, returning again in spring. In its usual wild haunts it attracts attention by continually flitting from the highest point of some bush, or tall weed, such as a thistle, to that of another at a little distance, occasionally alighting on the ground to pick up some insect which it has noticed, and from time to time uttering its peculiar call-note, which resembles the words, "Tsak, tsak, tsak." Sometimes, however, it may be seen perched on prominent mole-hills and large stones, and the highest parts of the cope of dry-stone dykes are also favourite resting-places.

---

[1] The Wheatear is often erroneously called the "Stanechacker" in Berwickshire.

Mr. Hardy, in his interesting papers on the local migration of birds,[1] mentions Penmanshiel Moor, Cockburnspath Cove, the heights above Redheugh, Piperton Hill or Earnslaw, Oldcambus Dean, Dowlaw Dean, and Ewieside,[2] as some of the usual haunts and breeding-places of the Stonechat. Selby states, in a paper "On Birds observed in the neighbourhood of Coldingham in April, and St. Abb's Head in June, 1833,"[3] that "upon the stony hills around Coldingham Loch, the Whin- and Stonechats were observed where whin or furze prevailed." It is still to be found in the last-mentioned locality, and it also frequents the hills around Abbey St. Bathans. Dr. Stuart records that it was found in considerable numbers about Gordon Bog when the Berwickshire Naturalist Club visited that place on the 30th of June 1879.[4]

I lately saw a beautiful specimen which had been taken with bird-lime by the gardener at Abbey St. Bathans, near the Lady's Pocket Wood there, in the end of April 1885. It is sometimes observed at Greenhope on Ellemford Farm, and occasionally on Cockburn Law,

> The restless Stonechat all day long is heard.[5]

The Rev. George Cook, Longformacus, informs me that it is often seen near Whitchester. It is sometimes observed in

---

[1] See *Hist. Ber. Nat. Club*, vols. vii., viii., ix., and x.

[2] Ewieside is one of the heights of the Lammermuirs in the neighbourhood of Oldcambus, and is mentioned by Walter Chisholm, a Berwickshire poet, in his lines on "The Pease Glen":—

> Down sank the red sun in his glory and splendour,
> Afar in the bowers of the cloud-curtained west,
> And soft, floating purple light, mellow and tender,
> Illumined wild Ewieside's heath-covered crest.

Walter Chisholm was born at Easter Harelaw, near Chirnside, in 1856, spent his boyhood at Redheugh, where his father was shepherd, and died at Dowlaw, near Fast Castle, in the twenty-first year of his age. His poems have been published in a little book edited by Mr. William Cairns, formerly of Oldcambus.

[3] *Hist. Ber. Nat. Club*, vol. i. p. 22.

[4] *Ibid.* vol. ix. p. 230.

[5] Wordsworth, *Evening Walk*.

autumn at the side of the Tweed near Paxton Toll. The food of this bird consists of worms and insects, varied with small seeds of several kinds. With regard to its song, Mr. Seebohm says:—" In spring, when all nature seems reviving under the cheerful beams of a brighter sun, the Stonechat's melody is amongst the first to inform us of the change of season. It is the first music heard on the uplands except indeed that of the Skylark. Long before the Meadow Pipits are in song, or the Buntings chant their monotonous music, the little Stonechat may be heard to pour forth his cheering notes. The little creature starts, may be from a spray of broom, which rebounds and quivers as he leaves it, and, fluttering in the air, he utters his music and retires to his perch again. His song is, like his flight, short and irregular." [1]

It usually builds its nest [2] towards the end of April or beginning of May, placing it on the ground at the bottom of some whin, juniper, or other moorland bush, and sometimes in a tuft of heather. The nest is composed of dry grass, moss, hair, and a few feathers, and the eggs, which are generally five or six in number, are of a pale bluish green colour, closely mottled with fine pale reddish brown specks. This species may be easily recognised, amongst the little birds of our wild deans and moorlands, by its nearly black head and throat, and the white patch on each side of its neck.

---

[1] Seebohm, *History of British Birds*, vol. i. p. 319.

[2] In some parts of Scotland the "Stanechacker" is exempted from the woes and pains of harrying, in consequence of a malediction which the bird itself is fancifully supposed to be always pronouncing—

   Stane-chack!
   Deevil tak!
    They wha harry my nest
    Will never rest,
    Will meet the pest!
   De'il brack their lang back,
   Wha my eggs wad tak, tak!

R. CHAMBERS, *Popular Rhymes of Scotland*, new ed., p. 189.

THE REDSTART.

REDTAIL, FIRETAIL.

Phœnicura ruticilla.

> The sun, the season, in each thing
> Revives new pleasures, the sweet Spring
> Hath put to flight the Winter keen,
> To glad our lovely Summer queen.
>
> UNCERTAIN—ABOUT 1590.

THE Redstart, which is one of our summer migrants, generally arrives in Berwickshire from the third week of April to the second week of May, when it may be occasionally seen flitting along the edges of woods near road-sides, and after a short flight alighting, with a jerk of its tail, on hedges, walls, and the lower branches of tall trees. It is a rather conspicuous little bird, the bright colours of the male, with the pure white spot on his forehead and the red feathers under his tail, being very attractive. The males arrive a few days before the females, which is the case with most of our migratory warblers.

About twenty or five-and-twenty years ago, the Redstart was much rarer in the county than it is at present, and it is one of those birds which have increased in numbers of late years, as mentioned by Dr. Robson-Scott, in his Anniversary Address as President of the Berwickshire Naturalists' Club, on the 24th of September 1874. It is now an annual visitor to various districts of the county, where, in former times, it was not seen. Amongst a few MS. notes on birds by the late celebrated Dr. Johnston of Berwick, written about thirty-five years ago, which his daughter, Mrs. Barwell-Carter of the Anchorage, Berwick, has been so good as to place at my disposal, I find that the Redstart is noted as "not common." The first note of the appearance of this bird in Berwickshire by Mr. Hardy, is on the 2nd of June 1842, when he saw it about Blythe Edge in Lauderdale, and building in the dykes about Girrick in the Parish of Nenthorn. On the 10th of May 1850, he records it as seen at the forester's cottage in Penmanshiel Wood, where he says it used to build, and ten days later he finds the nest in the outhouse there.[1]

Mr. Romanes of Harryburn has informed me that shortly after the last-mentioned date, a pair, which were amongst the first observed near Lauder, built their nest in a hole of the scaffolding of a new conservatory which Lord Lauderdale was then erecting at Thirlestane Castle. The Redstart was noticed for the first time at Paxton about 1867, when it attracted the attention of the workmen in Paxton Woods by its unusual appearance. It is now seen in every district of the county. Colonel Brown informs me that it is plentiful about Longformacus, and the Rev. George Cook tells me that Mr. Smith of Whitchester discovered a nest in the summer of 1886 in his grounds near the Dye. On the 15th of July 1886 I heard its call in the beautiful birch woods near the junction of the Dye and the Watch, and I have often noticed it about Abbey St. Bathans and Retreat Woods. Mr. Ferguson says it is common in Duns Castle Woods, where there are piles of stones for the protection of the tree roots, and it was noted as plentiful in the woods and plantations about Langton, Ninewells, and Blackadder, by Mr. W. Evans on the 6th of June 1886.[2]

---

[1] Mr. Hardy's *MS. Notes.*
[2] Mr. W. Evans, Edinburgh, in letter dated 18th Dec. 1886.

The song of the Redstart is pleasing, being somewhat like that of the Wren, though not so loud; and its call-note, which it often repeats, is like the words "Weet-tit-tit." It usually places its nest in a hole of a wall or tree, but it does not confine itself entirely to these situations, the nest being sometimes built in odd places, after the manner of the Spotted Flycatcher. As an instance of this, it may be mentioned that in the summer of 1886, a pair built their nest inside a wooden pump standing near the gardener's house at Paxton. The pump had not been used for some time, and the birds got in and out by the hole where the handle passed through the outside woodwork. The nest is generally constructed of moss and dry grass, lined with feathers and hair, with the addition of wool occasionally. The eggs, which are five or six in number, are very like those of the Hedge Sparrow, but are rather smaller, and of a paler blue colour. The food of the Redstart consists almost wholly of insects, and it generally leaves us for the south in August and September.

The male may be easily distinguished from all other Berwickshire birds, by his leaden grey back and head, the pure white spot on his forehead, and his chestnut-coloured breast and tail. The female has brownish plumage, with a reddish tail. Both the male and female attract attention by their ruddy colour when seen flitting about.

THE REDBREAST.

ROBIN, ROBINET, RUDDOCK.

Erithaca rubecula.

𝕿𝖍𝖊 𝕽𝖔𝖇𝖎𝖓, 𝕿𝖍𝖊 𝕽𝖔𝖇𝖎𝖓 𝕽𝖊𝖉𝖇𝖗𝖊𝖆𝖘𝖙.

And thou the bird whom men love best,
The pious bird with scarlet breast,
Our little English Robin;
The bird that comes about our doors
When Winter winds are sobbing.
Art thou the Peter of Norway boors,
Their Thomas in Finland,
And Russia far inland?
The bird who, by some name or other,
All men who know thee call their brother,
The darling of children and men.
                                WORDSWORTH.

THE Robin Redbreast is one of our most favourite birds, not only on account of its associations with our earliest years, but also from its fearless and confiding habits[1] in winter, and its cheerful song in early spring and autumn.

It is to a great extent a migratory bird in Berwickshire; a large proportion of those which are bred in this county, being joined by migrating Redbreasts from the north of Europe in September and October, shortly afterwards depart with them towards the south,[2] and leave only a few, which

---

[1] The earliest Scottish poet who mentions the Redbreast is Holland in *The Houlat*, written about 1453, and, apparently from its familiar disposition, he calls it the "henis-man," or family servant.—Sibbald, *Chron. Scot. Poet.* vol. ii. p. 369.

[2] Mr. Seebohm says that in those districts where the winters are severe, it migrates southwards in autumn to South Europe, North Africa, Palestine, and the cultivated districts of North-West Turkestan.—Seebohm, *Hist. Brit. Birds*, vol. i. p. 262.

remain with us through the winter. In autumn and spring it is pretty regularly seen on migration at the lighthouses on the coasts of England and Scotland, including those at the Farne Islands and the Isle of May.[1] Fewer are seen at the lighthouses in spring than in autumn, but this may probably be accounted for by the destruction of vast numbers while on migration, and before their return in spring, by birds of prey, and likewise by bird-catchers. In Yarrell's *British Birds* we are told that the latter take great numbers in France and Italy:—" During the southward migration of the Redbreast, it is caught in very large numbers for the table, and in autumn the bird markets of most towns in Southern France and Italy are generally well supplied with this species, which, among many others, passes indifferently under the name of *Beccafico*. Mr. Sclater observed more Redbreasts than any other birds in the Roman markets; and Waterton tells us that on characteristically expressing his regret at seeing so many in a stall there, the dealer assured him that if he took home a dozen for his dinner on that day, he would come back for two dozen on the morrow. These birds are usually taken in snares, or by limed twigs set round a captive Little Owl, which serves to attract the victims."[2]

The song of the Redbreast is sweet and plaintive, and is continued from early spring until winter.

> Each season in its turn he hails; he hails
> Perched on the naked tree, Spring's earliest buds:

---

[1] The Redbreast was observed on migration in *Autumn* 1880.—At Isle of May 18th Sep. and 13th Oct. *Autumn* 1881.—Isle of May, 22nd and 23d Sep., and 21st Oct. *Spring* 1882.—Isle of May, in April, and spring generally. *Autumn* 1882.—Isle of May, 18th Aug.; vast rush on 12th Oct., also 18th to 23rd Oct.; Farnes, 6th to 25th Oct. *Spring* 1883.—Isle of May, rush on 26th April. *Autumn* 1883. —Isle of May, rush on 15th and 22nd Sep. *Spring* 1884.—Isle of May, rush on 28th, 29th, and 30th April; Farnes, 11th March and 28th April. *Autumn* 1884.—Isle of May, 3rd Nov., a few.—*Reports on Migration of Birds*, 1879 to 1884.

[2] Yarrell, *Brit. Birds*, vol. i. pp. 310, 311.

> At morn, at chilly eve, when the March sun
> Sinks with a wintry tinge, and Hesper sheds
> A frosty light, he ceases not his strain;
> And when staid Autumn walks with rustling tread
> He mourns the falling leaf.
> <div align="right">GRAHAME, <i>Birds of Scotland</i>.</div>

The notes have a very pleasing effect as they fall upon the ear from the topmost twig of some tall bush or tree in an autumn evening, when the little songster

> . . . Pours his sweetest strains
> To charm the lingering day.
> TANNAHILL.

The fondness of the Robin for an elevated perch when singing did not escape the notice of our national poet, who, mourning over the departure of summer, says—

> Nae mair the flower in field or meadow springs:
> Nae mair the grove wi' airy concert rings,
> Except perhaps the Robin's whistling glee,
> Proud o' the height o' some bit half-lang tree;[1]

and the following fragment of an ancient song, written about 1575, shows that the autumnal strains of this bird had even then attracted attention—

> Robyn Readbrest with his noates,
>   Singing aloft in the quere,
> Warneth to get you frese coates,
>   For winter then draweth nere.

It was at one time a popular belief in the county that if the Robin were heard singing much, it was a sign of rain; and, alluding to this, Captain Bell, Linthill, near Eyemouth, in his MS. Diary of the Weather for 1802, records, on April 4th of that year:—"Robin Redbreast singing a good deal to-day—a sign of rain." Mr. Hardy, writing on the 23rd of October 1834, says :—" Rainy day. In the midst of the rain the Redbreast is singing as merrily as if it were summer-tide;" and, quoting from "Warwick," he adds:—'I am sent to the ant to learn industry, to the dove to learn

---

[1] Burns: " The Brigs o' Ayr."

innocency, to the serpent to learn wisdom, and why not to the Redbreast, who chaunts it as delightfully in winter as in summer, to learn equanimity and patience?'"[1] He also remarks that a popular saying, sometimes heard in the county before an impending change, is that "the Robins are too muckle aboot the doors for good weather;" and that it is believed that if they sit high and sing in harvest, it is a sign of good harvest weather. He likewise mentions that boys used to believe that they follow them into the woods for the purpose of intimating any danger that may waylay them, and sometimes this belief was so impressed upon them, that they would take to their heels if the birds approached too near them.[2] In some districts of the county the Robin is looked upon by the country people as a hallowed bird, and very few boys will kill one; it being said that if they do so its spirit may some day return, and seek the blood of the slayer. It is also said that it was the only bird which ventured near the Cross, and that the blood of our Saviour fell on its breast, which has remained red ever since.[3]

The old nursery ballad of the "Babes in the Wood,"[4] with which we have all been familiar in our childhood, contains an allusion to a kindly popular superstition regarding this bird piously covering the bodies of the dead with leaves :—

Thus wandered these poor innocents, till deathe did end their grief;
In one another's arms they dyed as wanting due relief:
No burial this pretty pair of any man receives,
Till Robin-Redbreast piously did cover them with leaves.

---

[1] Mr. Hardy's *MS. Notes.*   [2] *Ibid.*
[3] Mr. W. Lockie, Spottiswoode, in a letter dated 29th January 1887. [For superstitions regarding the Robin, see Swainson's *Folk Lore of British Birds*, and Rolland's *Faune Populaire de la France.*]
[4] It is mentioned in Ritson's *Ancient Songs and Ballads* that *The Children in the Wood; or The Norfolk Gentleman's Last Will and Testament*, appears to have been written in 1595, being entered in that year in the Stationers' Books, but the oldest edition now known in print is that entitled *The Cruel Uncle.* 12mo, 1670.

The food of the Redbreast in spring and summer consists chiefly of worms and insects, but, as the season advances, berries and garden fruit are also to some extent eaten by it. When digging is going on in the garden, it may often be seen flying down from its perch on some adjoining bush, and with confidence alighting within a few feet of the gardener, to feed upon any small worms which the spade may turn up. During winter it leaves the woods and plantations, and frequents farm-yards, stables, cattle-sheds, ashpits, and the like, to pick up what food it can find.[1] In apparent allusion to this habit, Sir David Lindsay says in *The Complaynt of Scotland*:—the "Robeen and the litil Vran var hamely in vyntir."[2] With the occurrence of the first snow-storm it draws closer to our doors and windows, and some morning soon afterwards, when the snow lies thickly over the ground, and the frost is severe, we find that

> . . . The old Robin has come
> To remind us with tip-tapping bill,
> That his morning repast of the delicate crumb,
> Should be spread for him now on the sill.
>
> Cook.

It has been observed that the same Robin sometimes comes for three or four years in succession to a particular window, to be fed with crumbs. This was noticed at a window of my father's house at Salton, in East-Lothian, when I was a boy, during the severe winters which occurred when our soldiers were in the Crimea, the bird being recognised by its having a twisted foot.[3]

---

[1] Mr. Hardy writes that he has seen it frequenting the caverns on the sea-coast in winter.—*MS. Notes.*

[2] *The Complaynt of Scotland*, p. 60.

[3] Mr. Hewitson mentions a similar instance in his *Eggs of British Birds*, vol. i. p. 100; and so does Mr. St. John in his *Natural History and Sport in Moray*, p. 116. Mr. Hardy records in his *MS. Notes*, that a Robin frequented the window of Mrs. Sinclair's house in Eyemouth, for several winters in succession, about 1844.

Our favourite is very pugnacious, and immediately attacks any small bird, including those of its own kind, which may venture to intrude upon its favourite haunt. While visiting lonely and retired parts of woods and plantations during summer and autumn, and waiting for a few minutes quietly there, I have often been surprised at the appearance of a Redbreast on the scene. This habit of the bird appears to be alluded to in the *Flyting of Polwart*[1] *and Montgomery*, where the former says to the latter—

> Into the land where thou was born,
>   I read but nought that it was skant
> Of cattel, clething, and of corn,
>   Where wealth and well-fair baith doth want.
> Now, tade-face, take this for no tant,
>   I hear your housing is right fair,
> Where howlring howlets ay doth haunt,
>   With Robin-Redbreast but repair.[2]

It builds its nest, which is composed of moss, dried grass, and dead leaves, lined with hair, early in spring; generally placing it amongst the herbage of a bank by the side of a road or ditch, or in a hole in an old wall, but it is found in various other positions. A remarkable situation for the nest is mentioned in the *Berwickshire Advertiser* of the 23rd of May 1845, by a correspondent in Coldstream :—" A pair of Robins, seeking for a place to build, fixed on the sleeve of a coat hanging in a shed at The Hirsel gardens. The hen bird is now sitting on six eggs, and, undismayed, seems to

---

[1] It would appear that this was Sir Patrick Home, the fifth of Polwart, and sixth in lineal descent from the first Sir David Home of Wedderburn, who died in 1469. Sir Patrick Home, the fifth of Polwart, died in 1689.—*Genealogical Table and Pedigrees in the Case of the Earldom of Marchmont*, 1820. See also *The Men of the Merse*, by Archibald Campbell Swinton, Younger of Kimmerghame, Edin. 1858, p. 28.

[2] Sibbald, *Chronicle of Scottish Poetry*, James VI., 1567-1603, Edin. 1802, vol. iii. p. 393. Mr. Hardy writes to me :—" I think the verse of Polwart is a satirical description of Montgomery's dwelling—the haunt of Howlets and Robins, which frequent dreary places. ' But repair '—it cannot be amended—there is no help for it."

welcome the curiosity of the numerous visitors to the singular domicile. The circumstance is rendered even more strange by the shed having been newly roofed during the time of her sitting." The eggs, which are generally four or five in number, though they sometimes reach six and seven, are white, thickly spotted with reddish yellow or grey.

The Redbreast is not subject to much variation in plumage. Mr. Romanes, of Harryburn, Lauder, has kindly informed me that two young Robins of a yellow colour were observed frequenting his garden in July 1886. One of them was caught, and preserved for the Museum of Science and Art in Edinburgh, where it may now be seen.

THE WHITETHROAT.

WHEY-BEARD, WHEETIE-WHY, NETTLE-CREEPER, CHURR, MUFF, MUFFET, MUFTY, CHARLY-MUFTY, BEARDY, BLETHERING-TAM, WHATTIE, WHISKEY.

Sylvia rufa.

𝔚𝔥𝔞𝔱𝔶-𝔴𝔥𝔢𝔶-𝔟𝔢𝔞𝔯𝔡, 𝔍𝔢𝔫𝔫𝔶-𝔠𝔲𝔱-𝔱𝔥𝔯𝔬𝔞𝔱, 𝔚𝔥𝔲𝔰𝔨𝔦𝔢,
𝔚𝔥𝔲𝔰𝔨𝔦𝔢-𝔴𝔥𝔢𝔶-𝔟𝔢𝔞𝔯𝔡, 𝔏𝔞𝔡𝔶-𝔩𝔦𝔫𝔱𝔶-𝔴𝔥𝔦𝔱𝔢.[1]

> *The happy Whitethroat on the swaying bough,*
> *Rocked on the impulse of the gadding wind*
> *That ushers in the showers of April, now*
> *Carols right joyously.*
>   CLARE.

THE well-known notes of this lively summer visitor are generally heard for the first time in the season early in May, coming from some bush or thicket of brambles and dog-roses by the roadside, where the male may be often seen perched, with erect crest and puffed-out throat, warbling forth his song in a hurried and restless manner. Occasionally he may be observed flying a little distance up into the air singing, and then returning to his perch, where, with many gesticulations, he continues his garrulous song.

The Whitethroat is a common species in Berwickshire during summer, and its nest is often found by boys; who seldom scruple to harry it, from a prejudice against the

---

[1] Mr. Hardy says that this name applies to the young after they are fully fledged, because they are in appearance and attitude not unlike Linnets.—*Hist. Ber. Nat. Club,* vol. x. p. 561.

"Whaty" on account of their belief that it sucks the eggs of other small birds. One of its popular names in the county —"Jenny-cut-throat"—would seem to indicate that it is thought to be guilty of even worse deeds than sucking eggs. It is unfortunate for the poor Whitethroat that it should be subjected to persecution from an erroneous idea that it sucks other birds' eggs or kills their young, for it is quite innocent of all such misdeeds, and lives entirely upon insects and small fruits. To obtain the latter it often visits our gardens in July, August, and September, when it may be seen amongst the currant and raspberry bushes. Mr. Hardy has noted that it is very fond of eating the berries of the *Daphne Mezereum* in August when they are ripe.[1]

It is a sprightly bird, and the traveller along our roads and bye-ways sees it now disappearing through the hedge, and again perched on some high twig at a little distance along the pathway, babbling out its song. I have pleasing remembrances of trout-fishing mornings on the Whitadder, in early summer, when, in many of the plantations on the banks of the stream, which abound with wild-flowers,—

> The sporting Whitethroat, on some twig's end borne,
> Pour'd hymns to freedom and the rising morn.
> 										BLOOMFIELD, *Spring.*

Although not a shy bird, it sometimes likes to hide itself amongst the thickest herbage, where its frequently repeated notes of "Hweet-hweet-hweet," and "Cha-cha-cha," indicate its presence.

It generally leaves us for the south in August and September. The nest is found in various positions, such as amongst tangled herbage by the side of a road or stream, in brambles or whins, in nettles or other rank weeds, and the like. It is loosely built of dry grass, and lined with finer

---

[1] Mr. Hardy's *MS. Notes.*

THE WHITETHROAT.

bents and horse-hair. The eggs, which are from four to six in number, are generally greenish white, speckled with olive green or light brown.

No specimen of the Lesser Whitethroat has as yet (March 1888) been obtained in Berwickshire.

THE BLACKCAP.

MOCK NIGHTINGALE, BLACK-HEADED PEGGY.

Sylvia Atricapilla.

> *The Blackcaps in an orchard met,*
> *Praising the berries while they ate ;*
> *The Finch that flew her beak to whet*
> *Before she joined them on the tree.*
>
> JEAN INGELOW.

THIS melodious summer visitor generally arrives in Berwickshire from the south between the third week of April and the third week of May, and leaves us in September.

It is found during summer in most of the deciduous plantations in the county, and especially those in policy grounds and deans, where the underwood is dense and intermingled with brambles and dog-roses. The Pease Dean and the woods about Paxton and Whitehall are some of its favourite resorts. It is one of the small birds which are apparently increasing in numbers, and occupying new districts where they were not previously observed.[1]

As a boy in Haddingtonshire, about thirty years ago, I was constantly on the outlook for birds and their nests in the district of Salton, and yet I never saw the Blackcap or heard its song. Dr. Turnbull, in his *Birds of East-Lothian*, 1866, mentions it as rare, but I am informed by my friend,

---

[1] The Rev. George Cook, Longformacus, informs me that he saw a Blackcap in the manse garden there in the summer of 1886 for the first time. It had not previously been observed in that neighbourhood.

Mr. W. Evans, that it is now comparatively plentiful in that county. There has been a considerable increase in its numbers about Paxton since 1870. Mr. Kelly, writing in 1876, says that it was not observed in Lauderdale before that year, when one was found dead in the grounds of Thirlestane Castle.[1] The first notice which I have of the bird's actual occurrence in Berwickshire[2] is by Mr. Hardy, who, writing on the 21st of May 1860, says:—" Blackcaps in many places in the Pease Dean, warbling richly; one came out and displayed itself in great agitation of song, its head feathers raised up and erect, its throat feathers disordered, its back curved—a rustic, ungainly-looking bird —and all the time singing earnestly, as could be seen by the working of its throat."[3]

The males usually come a few days before the females, and very shortly after their arrival they make the woods and deans resound with their delightful song, which is considered to be almost equal in compass to that of the Nightingale,

> Whose trembling notes steal out between
> The clustered leaves, herself unseen.
> MOORE.

The Blackcap is rather restless and shy, and keeps moving from branch to branch towards thicker covert on being observed. Its food consists of insects and fruit of various kinds, such as raspberries and currants, and I have observed that it is particularly fond of the fruit of the

---

[1] *Hist. Ber. Nat. Club*, vol viii. p. 143.
[2] It is mentioned by Selby in his "Report on the Ornithology of Berwickshire and District within the limits of the Berwickshire Naturalists' Club," in 1841, as among the rarer species of periodical summer visitors.— *Hist. Ber. Nat. Club*, vol. i. pp. 252-257. The district within the limits of the Club includes a large part of Northumberland, where the bird was probably a rare periodical summer visitor before it appeared in Berwickshire.
[3] Mr. Hardy's *MS. Notes.*

scarlet-berried elder in August, when the brilliant-coloured bunches hang temptingly upon the tree.

It generally selects a quiet, secluded spot in a plantation or wooded dean for its nest, which is built of dry grass lined with hair, and placed in a bush such as a sloe, briar, privet, or bramble, a few feet from the ground. It has nested every summer in Paxton Woods for the last fifteen years. The eggs are usually four or five in number, and vary considerably in colour, but are generally of a dirty white, clouded with yellowish brown, and marked with a few brown spots.

THE GARDEN WARBLER.

GREATER PETTYCHAPS, FAUVETTE, GARDEN FAUVETTE.

Sylvia salicaria.

> Trees have their music, for the birds they shield
> The pleasing tribute for protection yield.
> CRABBE.

THIS Warbler, like the last species, generally arrives in Berwickshire from its southern winter quarters, from the third week in April to the third week in May, and, like that bird, it appears to be increasing in numbers and extending to new districts. It was observed by me at Paxton, for the first time, on the 23rd of May 1879.

As it keeps itself closely concealed in the thickest foliage of the deciduous trees and underwood of our copses and deans, it is seldom seen; and in many cases its presence is only known to those who are acquainted with its song, and can distinguish its notes from those of the Blackcap, which they resemble.

The Garden Warbler is rather scarce in the county, but its notes may be heard in most of the woods which surround the houses of the landed proprietors, and it likewise frequents plantations throughout the Merse,[1] especially those which have plenty of low undergrowth in them.

---

[1] Mr. W. Evans has informed me that he saw it in the valley of the Whitadder in the summer of 1886, and in the woods of Whitehall in 1887.

THE GARDEN WARBLER.

A thickly-wooded dean, with a burn meandering at its bottom, is a favourite resort. It is not easily seen, for it usually conceals itself amongst dense foliage, and on the slightest alarm retires quietly to the thickest part of the covert.

Its song is mellow and sweet, and not so loud and irregular as that of the Blackcap, next to which it ranks as a songster. The two birds may be easily distinguished, for the Garden Warbler is without the black on the top of the head. It feeds on insects, fruit, and berries of various kinds, such as currants, blackberries, and elder-berries.

The nest, which is usually placed near the ground amongst brambles or briars, or other low bushes overgrown with herbage, is slightly built of withered grasses, and lined with a little horse-hair. The eggs are very like those of the Blackcap, being generally of a greyish white, slightly spotted with brown, and are usually four or five in number.

It leaves us in August and September for the south.

THE GOLDEN-CRESTED WREN.

GOLDEN-CRESTED KINGLET, GOLDEN-CROWNED WREN,
MARIGOLD FINCH.

Regulus cristatus.

𝕿𝖍𝖊 𝕸𝖎𝖑𝖑𝖊𝖗'𝖘 𝕿𝖍𝖚𝖒𝖇, 𝕿𝖍𝖊 𝖂𝖊𝖆𝖗𝖞 𝖔𝖗 𝖂𝖍𝖊𝖆𝖗𝖞.

> Lovely bird! with thy golden crown,
> A kind and tender nurse art thou,
> Making thy nest of moss and down,
> And hanging it on the bending bough.
> There, rocked by the wave of the zephyr's wings.
> Amid the green branches it lightly swings,
> And a few clustering leaves of the forest-tree
> Will serve to shelter thy cradle and thee;
> Concealing thee safely from every eye,
> Until danger and fear have pass'd thee by.
>
>   SIR WALTER SCOTT, *Minstrelsy of the Woods.*

ALTHOUGH this interesting and beautiful little bird is generally to be found in our fir woods all the year round, yet it is a partial migrant; for those which remain here during winter receive additions to their numbers from the south in spring, and many leave us for a warmer climate in autumn.[1] They are seen passing the lighthouses on the sea-coasts of England and Scotland, including those at the Farne Islands and the Isle of May, while on migration in

---

[1] In some districts of the county most of the Golden-Crested Wrens seem to have migrated southwards on the approach of the terrible winters of 1878-79, 1879-80, and 1880-81. Mr. Hardy, Oldcambus, writing on the effects of the winter of 1878-79, says that "the fir-tree tops were tenantless during the summer of 1879 as far as regards Golden-Crested Wrens."—*Hist. Ber. Nat. Club*, vol. ix. p. 129.

spring and autumn,[1] but more appear to be observed in autumn than in spring. The males have been noticed at the Isle of May to precede the females on migration by several weeks in spring, which is the case with many of our smaller migratory birds. The Golden-Crested Wren may be often seen frequenting the young spruce fir plantations in small flocks, in autumn and winter, where its faint call-note attracts attention before the bird is observed. Here it diligently searches amongst the evergreen branches for insects, flitting from tree to tree rapidly, and occasionally "keeking" underneath the boughs to see if any insects are hidden there. It may also sometimes be seen exploring hedges. During spring and summer it is dispersed in pairs through the woods, policy grounds, and gardens. It is very plentiful in Milne Graden grounds, where Miss Georgina Milne-Home has frequently discovered its beautiful nest. One which she showed to me in the summer of 1885, and which contained three eggs, was built on a branch of a young Scotch fir, close to the trunk, about five feet from the ground. The tree upon which it was placed was growing by the side of a walk within a few yards of the entrance to the kitchen court, where people were constantly passing in and out; but the nest was so well concealed that it could

---

[1] The Golden-Crested Wren was observed on migration in *Autumn* 1880.—At Isle of May, 8th and 15th Oct.; at Farne Islands, 12th, 13th, and 22nd Oct. *Autumn* 1881.—Isle of May, 27th Sep., 24th and 27th Oct.; Farne, throughout October. *Spring* 1882.—Isle of May, 9th March to 10th April. The males preceded the females by several weeks. *Autumn* 1882.—Isle of May, 27th Aug. to 12th Nov.; Farne, 8th Oct. Mr. Cordeaux remarks that this autumn (1882) will long be remembered amongst ornithologists for the extraordinary immigration of Goldcrests on the east coasts of England and Scotland. *Spring* 1883.—Isle of May, 1st to 13th April; Farne, 2nd April. *Autumn* 1883.—Isle of May, 10th Oct. *Spring* 1884.—Farne, end of March, and 28th April. *Autumn* 1884.—Isle of May, 31st Aug. to 27th Sep.; Farne, Aug. to Nov. *Spring* 1885.—Isle of May, 12th to 23rd April; Farne, 1st May. *Autumn* 1885.—Isle of May, 1st to 19th Sep.; Farne, 16th and 17th Oct.—From *Reports on the Migration of Birds*, 1879-1885. Mr. John Wilson, Welnage, Duns, has informed me that a fisherman told him that upon one occasion ten or twelve Golden-Crested Wrens alighted on his boat while off the coast of Coldingham.

THE GOLDEN-CRESTED WREN.

not be seen until a close inspection was made. Another, found in June 1886, was placed under the end of a bough of a large yew by the side of a walk in the flower garden. It was supported underneath the point of the branch by being interwoven with the small sprays there. Both nests were built of moss, and warmly lined with feathers. The eggs are generally from seven to ten in number, and are yellowish white, mottled near the larger end with light reddish brown.

This species may at once be distinguished from all our smaller birds, by the yellow streak on the top of its head. Its song, which may be heard in spring and early summer, is sweet and pleasing, though not very loud.

No specimen of the Fire-Crested Wren has as yet (1888) been obtained in Berwickshire.

THE CHIFFCHAFF.

LESSER PETTYCHAPS, LESSER WILLOW WREN, LEAST WILLOW WREN.

Phylloscopus collybita.

> *Ye curious chanters of the wood,*
> *That utter forth Dame Nature's lays.*
>
> SIR HENRY WOTTON, b. 1568.

SELBY, in his "Report on the Ornithology of Berwickshire and district within the limits of the Berwickshire Naturalists' Club,"[1] mentions the Chiffchaff as "among the rarer species" and "not common." On the 23rd of April 1874, Mr. Hardy, Oldcambus, records it as frequenting the tall young larches in the Pease Dean, and as "a bird never observed before in this vicinity."[2] In the beginning of May 1878, my friend, Mr. Arthur H. Evans, of Cambridge, while staying with me here, recognised the note of this species in the Old Heronry Wood at Paxton, and a week or two afterwards—on the 22nd of May—I heard the bird uttering its peculiar cry at the top of a high elm tree near the Avenue Bridge, and, after watching some time, had the satisfaction of seeing it come out of the middle of the tree and alight on a branch close to the spot where I stood. I had not previously observed it at Paxton. On the 22nd of June 1879, Mr. George Bolam discovered a nest of this bird in

---

[1] This Report was written in 1841. See *Hist. Ber. Nat. Club*, vol. i. p. 252.
[2] *Ibid.* vol. vii. p. 280.

the Old Heronry Wood, and kindly gave it to me for preservation. Since then the woods about Paxton have been visited by a few Chiffchaffs every season. Dr. Stuart noticed it at Chirnside Bridge, on the 27th of April 1881,[1] and Mr. Arthur H. Evans remarks (1881) that it is "spreading in the valley of the Eye."[2] Mr. Hardy informed me (July 28th, 1886) that it had been heard about Grantshouse and Dunglass that year, and Dr. Stuart told me about the same time that it now frequents the woods of Whitehall. In May 1886, the latter saw it at the Pistol Plantings, Broomhouse, Broomdykes, and Hutton Hall;[3] and in the same month I observed it at Burnhouses, and Oatleycleugh. Mr. Ferguson, Duns, wrote to me (June 24th, 1887) that "several pairs are in Duns Castle policies this year." It therefore appears to be now visiting the county in increased numbers.

The Chiffchaff generally arrives here between the third week of March and the third week of April, when its loud and peculiar note, which resembles the words "Chip-chap, chivy-chivy," may be heard amongst the tops of tall trees such as the oak, ash, and elm, and the little bird may be seen hopping about amongst the branches searching for insects. It cannot easily be distinguished by its appearance when on the top of a tree from the Willow Wren, but when it is in the hand its darker brown—almost black—legs, toes, and claws, as well as its shorter wings, enable the observer to identify it.

The nest is generally placed in a low bush, and is covered with a dome, the entrance being from the side. It is composed of dry grass, dead leaves, and moss, and is lined with feathers. The eggs are generally six or seven

---

[1] *Hist. Ber. Nat. Club*, vol. ix. p. 555. [2] *Ibid*. p. 555.
[3] *Ibid*. vol. xi. pp. 568, 569.

in number, and are white, thinly spotted with purplish brown.

The Chiffchaff leaves us for the south in September, when

> Gradual the woods their varied tints assume;
> The hawthorn reddens, and the rowan-tree
> Displays its ruby clusters.
>
> GRAHAME, *Rural Calendar.*

THE WILLOW WREN.

WILLOW WARBLER, GROUND WREN, HAY-BIRD, HUCK-MUCK.

Phylloscopus trochilus.

The Willy-Muff.

> *The House-Sparrow buildeth in the eaves,*
> *The Whitethroat and the Willow-bird*
> *From belfries in the curling leaves*
> *Make their rejoicing matins heard.*
>
> BUTLER, *An April Song.*

THE pleasing note of the Willow Wren, which is associated in our memories with the return of spring, is generally heard in this county for the first time in the season, about the third or fourth week of April.[1]

It is one of the most numerous of our summer visitors, and its joyous little song enlivens every part of our woods and plantations during the early summer months. The sylvan banks of our beautiful streams—the Whitadder, Blackadder, Leader, Dye, and Eye, appear to be favourite resorts of this small warbler, and there the sweet cadence of its strains often falls upon the ear of the angler as he casts his fly over the shady pools, or leisurely walks under the trees—

> In whose cool bowers, the birds with many a song,
> Do welcome with their choir the summer's queen.[2]

---

[1] Selby says that he has observed that the arrival of this bird may be expected with the first southerly wind as soon as the larch becomes visibly green. —*Hist. Ber. Nat. Club,* vol. ii. p. 71.

[2] Verses by "Jo. Davors, Esq.," in *The Complete Angler.*

The food of this species consists entirely of insects, such as *aphides*, and it searches the leaves and branches of the trees for these with restless activity.

Its nest is built on the ground, usually amongst the long grass near the margin of a wood, or by the side of a walk or ride through a plantation, and is formed of moss, dry grass, and dead leaves, lined with feathers. The eggs, which are six or seven in number, are white, spotted with light red.

The Willow Wren leaves us in August and September. It may be distinguished from the Wood Wren as described in the article on that bird.

THE WOOD WREN.

WOOD WARBLER, YELLOW WARBLER, LARGER WILLOW WREN,
YELLOW WILLOW WREN, GREEN WREN.

Phylloscopus sibilatrix.

> *Worship, ye that lovers be, this May,*
> *For of your bliss the calends are begun ;*
> *And sing with us, Away! Winter, away!*
> *Come, Summer, come! the sweet season and sun.*
>
> JAMES I., *The King's Quair*.[1]

THIS beautiful and interesting little bird generally arrives in Berwickshire between the third week of April and the third week of May, when it may be seen in some of the woods, poised amongst the newly opened and delicate yellowish green leaves of the beech,[2] singing its sweet and peculiar song, which resembles the words "Twee-twee-twee," repeated slowly at first, but afterwards quickly, and accompanied towards the end by a quivering of the wings.

It appears to be increasing in numbers, and extending to new districts in the county yearly. Mr. Hardy records its first appearance in the Pease Dean on the 8th of May 1873,[3]

---

[1] Written by James I. of Scotland in commendation of Lady Jane Beaufort (daughter of John Beaufort, Duke of Somerset, and grand-daughter of John of Gaunt), when he was in seclusion at Windsor Castle, and whom he afterwards married.

[2] Selby says that he has noticed that the arrival of the Wood Wren may be expected with the first bursting of the buds of the oak and the beech.—*Hist. Ber. Nat. Club*, vol. ii. p. 71.

[3] *Hist. Ber. Nat. Club*, vol. vii. p. 110.

and also its occurrence there on the 24th of April 1874.[1] In the middle of May of the latter year, one was observed by me at Wedderburn Castle, near Duns. This Warbler was not noticed at Paxton until the 18th of May 1879, when Mr. George Bolam drew my attention to it singing on a tree near the gamekeeper's house. Dr. Stuart saw it at Chirnside Bridge, on the 26th of April 1881, and a little later at Edrom, Harelawside, and Blackburnrig Wood.[2] On the 30th of May 1883, its song was heard amongst the trees at Stainrig, on the occasion of the meeting of the Berwickshire Naturalists' Club in that neighbourhood.[3] Mr. W. Evans, Edinburgh, wrote to me on the 18th of December 1886, that on the 7th of June of that year, he had found the Wood Wren at Ninewells, and Langton, and abundantly in the Pistol Plantation. On the 17th of May 1887, while driving to Abbey St. Bathans, I heard its song constantly repeated in the woods by the side of the public road near Burnhouses, and at Oatleycleugh. The food of this bird consists of insects and their larvæ, which it obtains among the leaves of trees.

The nest, which is placed on the ground in woods, under a tuft of grass or amongst herbage, is domed, and composed of moss, dried grass, and dead leaves, lined with fine grass and hair. It differs from that of the Willow Wren or the Chiffchaff, by never being lined with feathers. The eggs, which are six or seven in number, are white, thickly covered all over with small spots of dark reddish brown.

The Wood Wren may be distinguished from the Willow Wren and the Chiffchaff by the broad streak of bright sulphur yellow over its eyes, and the pure green of the upper part of its body. It leaves us in August and September for its southern winter quarters.

Hist. Ber. Nat. Club, vol. vii. p. 280.
[2] *Ibid.* vol. ix. pp. 559, 560.   [3] *Ibid.* vol. x. p. 251.

THE SEDGE WARBLER.

SEDGE-BIRD, SEDGE WREN.

Acrocephalus schœnobœnus.

𝕿𝖍𝖊 𝕾𝖈𝖔𝖙𝖈𝖍 𝕹𝖎𝖌𝖍𝖙𝖎𝖓𝖌𝖆𝖑𝖊.

Fixed in a white-thorn bush, its summer guest,
So low, e'en grass o'er-topped its tallest twig,
A Sedge-Bird built its little benty nest,
Close by the meadow pool and wooden brig,
Where schoolboys every morn and eve did pass,
In seeking nests, and finding, deeply skilled,
Searching each bush and taller clump of grass,
Where'er was likelihood of bird to build.
Yet did she hide her habitation long,
And keep her little brood from danger's eye,
Hidden as secret as a cricket's song,
Till they, well-fledged, o'er widest pools could fly :
Proving that Providence is ever nigh
To guard the simplest of her charge from wrong.

CLARE, *The Sedge-Bird's Nest.*

ALTHOUGH the Sedge Warbler frequents many suitable localities in Berwickshire regularly every summer, yet, owing to its habit of keeping itself concealed in the midst of thick bushes and tangled herbage, it is not so well known as some of our other warblers. It is a summer visitor, generally arriving here from the south between the first and the third week of May, and departing southwards again in September.

Its usual haunts are thickets of stunted willows[1] or

---

[1] *Salix cinerea*, the Grey Saugh. "In peat bogs on all our moors, on banks, in deans, and in hedges, it often forms a little thicket, especially in oozy ground, by the sides of our muirland or dean burns, and these are favourite resorts of our song birds."—Dr. Johnston's *Botany of the Eastern Borders*, p. 181.

other bushes, intermingled with rank vegetation by the sides of our bogs, ponds, ditches, and streams,

> Where bur-reeds and rushes by the water-side bloom,
> And the lang meadow-queen waves its snawy white plume.[1]

Long ago the extensive morass of Billy Mire, which was covered from end to end with luxuriant sedges and bog-reeds, with low grey willows here and there on its banks, would form a favourite resort of this bird, and during summer its song would be heard in every direction amongst the rank vegetation which bordered the deep moss-hags and black pools of the mire. It is also found at some distance from water in thick hedgerows and young plantations, where the trees are growing close together. The most remarkable characteristics of this Warbler are its song—which is loud, garrulous, and imitative of the notes of various birds, such as the House Sparrow and the Whitethroat—and its habit of frequently singing late in summer nights, which has apparently given rise to the local name of *Scotch Nightingale*. I may mention that I heard one singing as late as 10.30 P.M. in a hedge near Nabdean, in the parish of Hutton, about the end of June 1886. The song was continued for a considerable time, and sounded sweetly in the distance, the notes being like those of a Lark.

Although the Sedge Warbler usually keeps itself well hidden, it may sometimes be observed in the morning singing on the top of a bush, or even on the branch of a tree. I find from my note-book that, on the 20th of May 1886, I saw one sitting singing on a leafless branch of a young ash tree near Nabdean Mill Pond, at 9 A.M., and noticed that it puffed out its little throat, and kept its tail close down on its perch while it sung. Sweet mentions that it generally

---

[1] Dr. Henderson's *MS. Poems*.

begins its song with "Chit-chit, chiddy-chiddy-chiddy, chit-chit-chit."

Selby, in his "Report on the Ornithology of Berwickshire," in 1841, mentions this species as one of those which at that time did not abound to the extent it formerly used to do, and says he is inclined to attribute the diminution of its numbers "to the effect of an improved system of agriculture, which by draining and reclaiming the marshy spots and little tangled thickets of sallow, and other rough and rampant herbage that used to be so common in our fields, has destroyed its favourite and appropriate breeding-places."[1] Its occurrence in Dowlaw Dean and Lumsden Dean is noted by Mr. Hardy as early as the 20th of June 1843.[2]

It was frequently heard singing in the thick hedges by the side of the Tweed at Paxton in the summers of 1873 and 1874;[3] and Dr. Stuart of Chirnside records that its note was recognised in Gordon Bog on the 30th of June 1879.[4] Mr. Hardy gives the following localities as some of its usual haunts :—In willow thickets at Penmanshiel and Dowlaw; mosses on Coldingham Moor; Cockburnspath Tower Dean; and the Pease Mill;[5] willow scrubs at Redheugh; Lumsden Moss; and Oldcambus.[6] It was noticed by Miss Georgina Milne-Home at the side of the Tweed at Milne Graden, in the summer of 1886; and in June of the same year I observed it at Coldingham Loch, and at Mertoun. Mr. John Ferguson, Duns, informs me that several Sedge Warblers frequented the beds of reeds and willows at the upper end of the Lake at Duns Castle, in the summer of 1887, where they were heard singing every evening.

The food of this bird consists of various kinds of aquatic insects and small worms.

---

[1] *Hist. Ber. Nat. Club*, vol. i. p. 252.
[2] Mr. Hardy's *MS. Notes*.
[3] *Hist. Ber. Nat. Club*, vol. vii. p. 382.
[4] *Ibid.* vol. ix. p. 229.
[5] *Ibid.* vol. ix. p. 560.
[6] *Ibid.* vol. x. p. 559.

Its nest, which is generally composed of grass and moss lined with hairs, and sometimes with the tops of umbelliferous plants, is rather deep, and is usually placed in a low bush or amongst tangled herbage. The eggs are four or five in number, and of a pale yellowish brown, freckled all over with a darker shade of brown.

It is about the size of a Willow Wren, and may be distinguished from our other warblers of that size, by the broad streak of yellowish white over its eyes, and by the top of the head being streaked longitudinally with dark and light brown.

The Reed Warbler (*Acrocephalus streperus*) has not been observed in Berwickshire.

THE GRASSHOPPER WARBLER.

CRICKET BIRD.

Acrocephalus nævius.

> *The Summer loves not silence ; her great charm*
> *Is in the concourse of a thousand sounds :—*
> *The birds, the winds, the very earth herself*
> *Breathing with life at every bursting pore,*
> *And that low melody that comes*
> *I know not whence or how.*
> FABER.

THE Grasshopper Warbler is a summer visitor, generally arriving in Berwickshire in small numbers about the end of May. It is found in a few suitable localities throughout the county, but as it is very shy, and skulks in the thickest covert, such as young fir plantations, where the ground is covered with heather, long grass, and other rank herbage, it is rarely seen. The presence of the bird in its favourite haunts is, however, indicated by its monotonous trill, which resembles the note of the Grasshopper, though somewhat louder and more prolonged. It was heard by me for the first time on the evening of the 9th of June 1876, amongst some whin bushes in a plantation of silver firs on the farm of Nabdean, when the bird allowed me to approach quite close to the spot where it sat and uttered its peculiar song. It remained there only a few days. I did not hear the the note of the Grasshopper Warbler again until the 28th of May 1888, when my wife called my attention to it, as we were walking in my garden about half-past nine o'clock at night. On listening to the sound, we found it proceeded from a plantation of young firs immediately to the west of the garden, where the bushy trees about three or four feet in height, interspersed with numerous tufts of tall cocksfoot and other grasses, afforded such ample concealment to the bird that we could not obtain a view of it. Its trill was heard there every evening afterwards, until the end of the first week in June, when it ceased.

Having learned that Mr. John Barrie, son of Mr. Barrie, gamekeeper, Preston, had obtained a Grasshopper Warbler in that neighbourhood, in July 1888, I called on him at Preston, on the 12th of the following month, when he showed me the bird, which I found to be of this species. He has favoured me with the following notes regarding his experience of this Warbler:—" On the evening of the 6th of July 1888, I happened to be in a young plantation of Scotch firs, about five feet high, which grow among deep heather on the top of the hill between Hoardweel and Drakemyre Moors, when I heard a peculiar whirring sound. On the following night I went up earlier and waited, but nothing was heard until about eight o'clock, when the same sound again reached my ears, but not so loudly as on the previous evening. As the night advanced the noise grew louder, and although it seemed to proceed from a spot close to the place where I stood, I had to walk fully thirty yards before I saw the bird which produced it. The bird was sitting in the centre of a young Scotch fir, and I watched it begin its song, which it commenced in a low key, and which gradually rose higher until it reached a certain pitch, when it stopped. The bird did not feed between its songs; it simply hopped about, or, if disturbed, flew to a short distance and again began its note. I then shot the bird, in order to determine the species." As Mr. Barrie has informed me that he heard several other Grasshopper

Warblers in different parts of the plantation on the evening when he obtained the above-mentioned specimen, it is very probable that they nested there in the summer of 1888.

This bird is recorded by Dr. Stuart as having occurred in 1880 at Whitehall, in the parish of Chirnside; also at Hammerhall, in the adjoining parish of Buncle; and in former years at Ninewells, near Chirnside.[1] Mr. Hardy mentions that its note was frequently heard about Penmanshiel[2] and the Pease Dean[3] long ago.

"The note," says Stevenson in his *Birds of Norfolk*, "if once heard, can never be afterwards mistaken for the sound of a grasshopper or cricket, however striking the resemblance; besides, the length of time for which it is continued, provided the bird be not disturbed, is much greater. Thus, on one occasion, while watching some pike lines by the margin of a deep pool, I heard the trill of the Grasshopper Warbler emitted from a neighbouring hedge for at least twenty minutes, during which time the bird appeared to have been sitting on the same spot."[4] Yarrell remarks that "in the more marshy parts of England, where the chirping of grasshoppers and crickets is not a very common sound, this bird has long been known as the *Reeler*, from the resemblance of its song to the reel used, even at the beginning of the present century, by the hand-spinners of wool. But this kind of reel being now dumb in such districts, the country folks of the present day connect the name with the reel used by fishermen as being that most familiar to them."[5] Seebohm describes the note of this bird as "exactly resembling the note of the grasshopper, except that it is slightly louder, not quite so shrill, and somewhat steadier and more prolonged."[6]

---

[1] *Hist. Ber. Nat. Club*, vol. ix. p. 229.
[2] *Ibid.* vol. vii. p. 511.   [3] *Ibid.* vol. ix. p. 229.
[4] *Birds of Norfolk*, vol. i. p. 104.   [5] Yarrell's *British Birds*, vol. i. p. 385.
[6] Seebohm's *British Birds*, 1883, vol. i. p. 341.

THE GRASSHOPPER WARBLER.

The food of this warbler consists of insects of various kinds.

The nest, which is generally formed of coarse bents and moss, lined with finer materials, is placed near the ground in a tuft of long grass, and is very difficult to find. The eggs are from four to seven in number, and are pinkish white, spotted all over with reddish brown.

The Grasshopper Warbler is somewhat larger than the Sedge Warbler, and the upper parts are greenish brown, the middle of each feather being darker; the lower parts are pale brown. The length of the bird is five and a half inches.

THE HEDGE SPARROW.

HEDGE WARBLER, DUNNOCK, DICK DUNNOCK, SHUFFLE-WING.

Accentor modularis.

𝕮𝖍𝖊 𝕳𝖊𝖒𝖕𝖎𝖊, 𝕮𝖍𝖊 𝕳𝖊𝖒𝖕 𝕾𝖕𝖆𝖗𝖗𝖔𝖜.

> *The sooty-plumed Hedge Sparrow frequent acts*
> *The foster-mother, warming into life*
> *The youngling destined to supplant her own.*
>
> GRAHAME, *The Birds of Scotland.*

THIS well-known, modest, and unobtrusive bird is resident in the county throughout the year, and frequents our hedge-rows, gardens, and woods, during spring, summer, and the early part of autumn. Towards the approach of winter it draws nearer to the neighbourhood of houses, farm buildings, and villages, where it picks up such scanty food as it can find in the shape of small seeds, crumbs of bread, and the like. It is one of the few birds which do not desert us even in the severest snow-storms, such as that of January 1881, when it was seen coming to our windows in company with House Sparrows, Blackbirds, Robins, Chaffinches, and Titmice, to be fed on the crumbs from our tables.

> And who can grudge so small a grace
> To suppliants, natives of the place?
>
> COWPER.

The Hedge Sparrow commences to sing very early in the year, its short and sweet song being frequently heard in

February.[1] During the fearful snow-storm of the 2nd and 3rd of March 1886, when everything was covered up with drifting snow, and the roads blown up in some places to to the depth of eight or ten feet, a pair of Hedge Sparrows sheltered themselves in my greenhouse for several days. As soon as the weather moderated, they left the shelter which they had enjoyed, and on the afternoon of the 5th of March I heard the male singing sweetly to his mate on an ash tree close by, although the snow-drifts still lay all over the district to the depth of four or five feet. The food of this bird consists of insects, worms, and seeds.

It generally commences to build its nest about the beginning of April, in some bush or hedge-row in the vicinity of our farm-steadings and villages. A Hempie's nest, with its five bright blue eggs, is a great attraction to boys, who love to look at and handle the eggs. It is often one of the first nests of the season which they discover. The Cuckoo sometimes selects the nest of this species in which to deposit an egg, with the result alluded to by Grahame in the lines above quoted; and Shakespeare, referring to this, says, in *King Lear*—

> The Hedge Sparrow fed the Cuckoo so long,
> That it had its head bit off by its young.

---

[1] Mr. Hardy records :—"1874, Feb. 16th—Hedge Sparrow singing."—*Hist. Ber. Nat. Club*, vol. vii. p. 277.

THE DIPPER.

WATER OUZEL, WATER CROW, WATER PIET, KINGFISHER.

Cinclus aquaticus.

𝕿𝖍𝖊 𝖂𝖆𝖎𝖙𝖊𝖗[1] 𝕮𝖗𝖆𝖜.

> *The Finch no more on pointed thistles feeds,*
> *Pecks the red leaves or crops the swelling seeds ;*
> *But Water Crows by cold brook margins play,*
> *Lave their dark plumage in the freezing spray,*
> *And, wanton, as from stone to stone they glide,*
> *Dive at their beckoning forms beneath the tide.*
>
>   LEYDEN, *Scenes of Infancy.*

THE Dipper is found on all our rapid streams and burns, to whose beauty it gives an additional charm at every season, but more especially during that delightful period of the year when the angler, with his rod and basket, wanders

> By shallow rivers, to whose falls
> Melodious birds sing madrigals.
>   MARLOW.

It may be then often seen sitting on a stone by the side or in the middle of the stream, bobbing its head, and jerking its tail up and down at intervals,[2] and displaying its beautiful white breast to great advantage. On being approached it flies off hurriedly along the course of the river, near the surface of the water, with rapid flight and quickly moving wings; generally uttering its usual note, which resembles the

---

[1] In Anglo-Saxon *water* is *wæter*. —Mr. Hardy's *MS. Notes.*

[2] This habit of bobbing up and down is alluded to by James Hogg when he says : " The factor's naig wantit a fore-fit shoe, an' was beckin like a Water Craw." —Dr. Johnston's *MS. Notes.*

words "Chit-chit," during its flight, and alighting at no great distance in a similar position to that which it occupied when it was disturbed.

The Dipper is always associated in my mind with pleasant trout-fishing days on the banks of the Whitadder and the Dye, when, as old Isaac hath it, " I could sit there quietly, and, looking in the water, see some fishes sport themselves in the silver streams, others leaping at flies of several shapes and colours; looking on the hills, I could behold them spotted with woods and groves; looking down the meadows could see, here a boy gathering lilies and lady-smocks, and there a girl cropping culverkeys and cowslips, all to make garlands for this present month of May."[1]

When looking for prey under the water, it does not dive into the stream from a little height like the Kingfisher, but either walks into or alights on the water, and then ducks under the surface, using its wings to swim under the water much in the same way as it uses them to fly in the air. This is a fact which has been thoroughly investigated and ascertained by ornithologists.[2] I have sometimes noticed it tumbling about in very shallow streams on the Whitadder, with its head down amongst the small stones at the bottom, when it appeared to be engaged in hunting for the numerous water insects which abound in such places. Its food consists chiefly of aquatic beetles and small fresh-water molluscs, many of which are known to be destructive to the spawn of fish.

The Dipper is an early breeder, generally commencing to build in March or in April. The nest, which is domed and well constructed, is somewhat like that of the Wren in external appearance; being chiefly composed of moss, lined with dead leaves, and having a hole in the side

---

[1] Isaac Walton, *The Complete Angler*.
[2] See Macgillivray, Yarrell, and Seebohm.

just large enough to admit the bird. It is usually placed in a bank overhanging the water, or under a moss-grown rock, by the side of a stream or burn, and is often so well concealed that it escapes the prying eyes of boys, who consider it a great prize.

> Up the bosky howe linn where the Water-Craw dooks,
> And laves his white breast 'mang the faim,
> We hae scrambled for 'oors mid the windins and nooks,
> Tae seek for his moss-theekit hame.
>
> THOMAS WATTS.

The same spot is often selected for the nest year after year, if the birds be not disturbed. While following the Hon. G. Hill's otter-hounds along the banks of the Whitadder, in company with Mr. John Clay, jun., on the 27th of May 1879, we discovered one placed in the angle of the buildings of Clarabad Mill, where the water rushes out after turning the water-wheel, and the miller told us that a pair of Dippers had built there every season, for at least ten years in succession. Mr. Thomas H. Ford, Duns, informs me that when he lived at Nisbet Mill on the Blackadder, upwards of fifteen years ago, the Water Crow was very numerous there, especially during snow-storms in winter, when it became comparatively tame. Its favourite breeding-place was in a hole under the arch of a small bridge across the mill-lade about thirty yards above the mill, which it was known to have occupied for forty years. At Abbey St. Bathans a nest was placed for twelve successive seasons near the water-wheel of the saw-mill; and at Longformacus House one is often built close under the footway of a wooden bridge which leads across the Dye to the garden. The rocky dean which runs to the Tweed near Paxton House, and through which a small rivulet meanders, is a favourite haunt of this bird, and here I have frequently found its nest under a moss-grown rock overhanging the stream. The eggs, which are five or six in number, are rather less than those of the Song Thrush, and are of a delicate pinkish white when the yolk is in them. They are pure white when blown. Two broods are frequently reared in the season.

The song of the Dipper is very pleasing, and often enlivens our streams and rivulets during the winter season, when snow is on the ground, and "the frozen rill's hoarse murmur scarce is heard." It may be said to be the latest as well as the earliest of our feathered songsters, its song being heard late in autumn, through winter, and early in spring, when—

> The ravaged fields, waste, colourless, and bleak,
> Retreating winter leaves, with angry frown,
> And, ling'ring on the distant snow-streaked hills,
> Displays the motley remnants of his reign.
>
> GRAHAME, *Rural Calendar.*

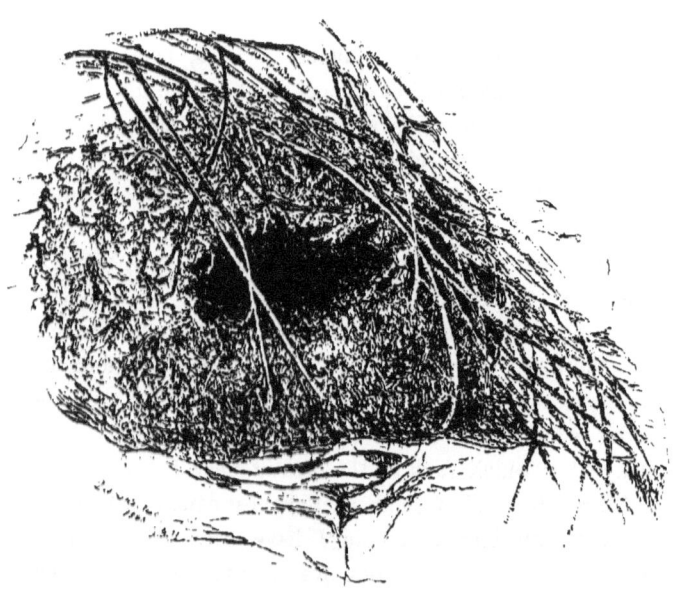

THE LONG-TAILED TITMOUSE.

BOTTLE TIT, BOTTLE TOM, LONG-TAILED MAG, HUCK-MUCK,
POKE-PUDDING, MUM-RUFFIN, MUFFLIN.

Acredula candata.

> *Birds of many dyes*
> *From tree to tree still faring to and fro.*
> Hood.

This interesting and various-coloured little bird, which remains with us throughout the year,[1] may be seen streaming through many of our tall woods and plantations[2] in small wandering flocks during the winter months,[3] keeping generally about the tops of the trees, and flying rapidly from branch to branch in search of insect food. The singular appearance and activity of the small creatures, with their long tails, as well as their constantly uttered notes, which resemble the words "Churchur" and "Twit-twit," seldom fail to attract attention, and help to enliven the woods during the period of the year

> When the trees stand like grim skeletons,
>   Nor spray nor leaflet can we find,
> Save where the gnarled and sombre oaks
> Defiant wave their straggling locks
>   Upon the wintry wind.
>   Walter Chisholm.

---

[1] The Long-Tailed Titmouse has been observed on migration at several of the lighthouses on the English coasts. I have not been able, however, to find any instance of its appearance at the Farne Islands or the Isle of May in the *Reports on the Migration of Birds*, 1879-86.

[2] The Pistol Plantation, near Blackadder, is a favourite resort of the Long-Tailed Titmouse.

[3] Colonel Brown of Longformacus has informed me that a few years ago, during severe weather in winter, a number of Long-Tailed Titmice came to a window of Cowdenknowes House, where they were fed by Mrs. Hope.

In spring they cease to be gregarious, and separate into pairs. About the middle of May they proceed to build their beautiful nest, which is oval in shape, with a small hole in the upper part of the side for the entrance of the bird. It is generally placed in the centre of a thick bush, where it is supported by twigs and branches, and is composed of green moss, woven throughout with wool and spiders' webs, and coated over the outside with pieces of grey lichen, like the nest of the Chaffinch. The inside is thickly lined with feathers, of which a nest mentioned by Macgillivray contained no fewer than 2379, chiefly belonging to the Pheasant, Wood Pigeon, Rook, and Partridge. The eggs, which are generally seven or eight in number, are white, with pale reddish spots. I discovered a nest of the Long-Tailed Titmouse in the policy grounds of Paxton, near the Avenue Bridge, in the summer of 1874.[1]

> But most of all it wins my admiration
> To view the structure of this little work,
> A bird's nest. Mark it well within, without,
> No tool had he that wrought, no knife to cut.
> No nail to fix, no bodkin to insert,
> No glue to join; his little beak was all.
> And yet how neatly finished. What nice hand,
> With every implement and means of art,
> And twenty years' apprenticeship to boot,
> Could make me such another?
>
>   HURDIS.

---

[1] Mr. W. Locke, teacher, Gateside School, Spottiswoode, has informed me that the nest of the Long-Tailed Titmouse is sometimes found about Carolside and Cowdenknowes. Mr. Watson, Duns, exhibited at a meeting of the Berwickshire Naturalists' Club, held at Belford on 28th July 1880, a beautiful nest of this bird found in a furze bush at Chapel, near Duns.

THE GREAT TITMOUSE.

OX-EYE, GREAT BLACK-HEADED TOMTIT, BLACK-CAP, SIT-YE-DOWN, SAW-SHARPER.

Parus major.

𝔒𝔵=𝔢'𝔢, 𝔅𝔢𝔢=𝔢𝔞𝔱𝔢𝔯, 𝔅𝔩𝔞𝔠𝔨=𝔥𝔢𝔞𝔡.

> *I have a prety tytmouse*
> *Come picking at my to,*
> *Gossuppe with you I purpose,*
> *To drink before I go.*
>
> SCRAP OF ANCIENT SONG, ABOUT 1575.

THE Great Titmouse is found in the plantations, strips, and wooded deans of the county during summer, and in winter when snow is on the ground it is often seen in the neighbourhood of our houses, farm buildings,[1] stackyards, and gardens.

It appears to be a partial migrant, for it occurs amongst the numbers of Titmice of various kinds which are seen passing the lighthouses on the coasts of England and Scotland in spring and autumn.[2]

---

[1] The tops and the eaves of stacks are favourite resorts. I noticed numbers of Great Titmice feeding on the stacks at Nabdean on 24th December 1886, when snow was lying on the ground to the depth of six or eight inches.

[2] Although there can be little doubt that the Great Titmouse migrates in numbers with the other Titmice which are observed on migration at the Farne Islands and the Isle of May in spring and autumn, the only record of its actual identification on migration at either of these places is at Farnes on 13th October 1882, when many old males were observed. Mr. Harvie-Brown, in a note in the "Sixth Report on the Migration of Birds," says with regard to Titmice : - " We wish our reporters would distinguish the species more exactly."

THE GREAT TITMOUSE.

It is a sprightly bird, and is always in motion, flitting about amongst the branches of trees and bushes, and searching the bark, buds, and twigs for insect food. While thus engaged it assumes a great variety of interesting attitudes, and may be sometimes seen clinging to a twig back downwards.

About Cockburnspath it goes under the name of the "*Bee-Eater*," and is said to be a great enemy to bee-hives in that neighbourhood, where it suddenly descends from the trees which surround the village gardens, and seizes the bees as they issue from the hives.[1] This habit has also been noticed in the village of Paxton, when snow is on the ground and the bees are tempted to leave the hives by bright sunshine.[2]

The cheery and oft-repeated spring notes of this bird are generally heard about the beginning of March for the first time in the season, and are continued until about the middle of May. They may be said to resemble the words "Tee-ta, tee-ta, tee-ta," and in some places are likened to the sound produced by the sharpening of a saw.

The nest, which is usually placed in the hole of a tree and sometimes of a wall, is composed of moss and feathers, with some hair. The eggs are from six to nine in number, and are white, spotted with light red.

---

[1] *Hist. Ber. Nat. Club*, vol. vii. p. 295 ; also vol. x. p. 563.
[2] My friend Mr. Ingram, of Belvoir Castle Gardens, informs me that during the severe snow-storm of 1886 several bright sunny days occurred, which induced his bees to come out of their hives, when they were immediately attacked by Great Tits, and many of them destroyed. He was so much annoyed at this destruction that he set some small steel traps baited with dead bees near the hives, and in the course of two days caught fifteen of the depredators.

THE COLE TITMOUSE.

COLE-TIT, COLEMOUSE.

Parus ater.

> The flowery glory of the year has fled
> To other climes, with many a bird, whose song
> Made jubilant the Earth all Summer long;
> And Nature mourns as mother for her dead.
>
> WALTER CHISHOLM, *November*.

IT is chiefly in the autumn and winter months that the movements of flocks of Titmice and Golden-Crested Wrens in our woods and plantations attract attention; for then the deciduous trees are bare, and the little birds can be more easily seen as they flit from tree to tree. Amongst these diminutive rovers the Cole Titmouse may generally be observed moving rapidly, and eagerly searching the branches and twigs for insect food. The pine woods and strips are, however, its favourite resort, and it may be found in them at every period of the year.[1] Its call-note, "If-hee, if-hee, if-hee," is generally heard proceeding from the top of some tall Scotch fir before the little creature can be discovered busily occupied amongst the bushy branches. It is said to feed partly on the seeds of the fir "top" or cone.

---

[1] The Cole Titmouse is a partial migrant like the other kinds of Titmice, and it helps to make up the flocks of these birds which are seen passing the lighthouses on the coasts of England and Scotland on migration in spring and autumn. —See *Reports on the Migration of Birds*, 1879-86.

THE COLE TITMOUSE.

The nest of the Cole Titmouse is generally built in a hole in a tree, or wall, or in the ground,[1] and is composed of moss, wool, hair, and feathers. The eggs, which are from six to eight in number, are white, spotted with light red, the spots being somewhat larger than those on the eggs of the Blue Titmouse.

This species is easily distinguished from the Marsh Titmouse, which it resembles, by the former having a white patch on the nape, and white spots on the wing coverts.

---

[1] Mr. William Evans discovered a Cole Titmouse's nest in a hole in the bank by the side of the small stream which flows past the gamekeeper's house at Paxton, on 12th June 1887. It had young.

THE MARSH TITMOUSE.

MARSH-TIT, BLACK-CAP, SMALLER OX-EYE, WILLOW-BITER, JOE BENT.

Parus palustris.

> *Pease! once I sung thee when autumn was glowing,*
> *When flow'rets were fading and Nature forlorn,*
> *When trees their sere leaves on the streamlet were throwing,*
> *And birds sat in silence on hazel and thorn;*
>
> *Now from the sunny south green spring returning,*
> *Blowing her life-giving breath o'er the Glen,*
> *Long buried beauties of Nature unurning,*
> *Calls me to strike my weak harp again.*
>
> <div align="right">WALTER CHISHOLM, <i>The Pease Glen.</i></div>

THE Pease Dean, a romantic and beautiful ravine in the parish of Cockburnspath, seems to be a favourite resort of the Marsh Titmouse, for it has been frequently observed there by Mr. Hardy, who noticed a pair building their nest in a decayed oak stump near the Pease Bridge, on the 4th of May 1882.[1]

It is by no means a common bird in Berwickshire, but is found in small numbers in many of the woods and plantations throughout the county. Like the Cole Titmouse, it remains with us throughout the year, and in its general habits it resembles that species, from which, however, it can readily be distinguished by the want of the white spot on

---

[1] *Hist. Ber. Nat. Club,* vol. ix. p. 562, and vol. x. p. 558.

the nape. I have occasionally seen it in the woods about Paxton. The food of the Marsh Titmouse consists of insects and various seeds and berries.

Its nest, which is made of moss and grass, and lined with willow down, is placed in a hole of a decayed tree near the ground, which it hollows out, removing the chips to some distance in its bill. The eggs are seven or eight in number, and are white, spotted with light red.

THE BLUE TITMOUSE.

TOM-TIT, BLUE-CAP, BLUE-BONNET, HICK-MALL, BILLY-BITER, OX-EYE, NUN, BLUE-MOPE.

Parus cæruleus.

𝕿𝖍𝖊 𝕯𝖝-𝖊'𝖊.

Lithest, gaudiest harlequin!
Prettiest tumbler ever seen,
Light of heart and light of limb,
What is now become of him?
Where is he that giddy sprite?
Blue-Cap with all his feathers bright,
Who was blest as bird could be;
Feeding on the apple tree,
Made such wanton spoil and rout,
Turning blossoms inside out;
Hung with head towards the ground,
Fluttered, perched into a round,
Bound himself, and then unbound.

WORDSWORTH.

THIS little bird is a great favourite with us on account of its beauty and familiarity. It accompanies the Robin to our window-sills during snowy weather in winter, to share the crumbs and meat[1] put out by kind hands for the starving birds; and we find it represented as the lover

---

[1] The Blue Titmouse is particularly fond of a bone or piece of suet to peck at in severe weather, when these are hung out at a window. The attitudes of the Tits, whilst feeding upon them, are most interesting and amusing. On 20th Jan. 1881, during the continuance of the great snow-storm, the only birds which came to our windows to be fed were Sparrows, Hedge Sparrows, Robins. Chaffinches, and Blue Titmice.

of the Wren in the following verse of an ancient Scottish song, called "Lennox's Love to Blantyre":—

> Where's the ring I gae thee,
> Of yellow gold sae fine?
> I gae't to my love Oxee,
> A true sweetheart o' mine.

The Blue Titmouse is very common, and is found in our woods, roadside strips, gardens, and hedgerows, nearly all the year round; where its active and amusing habits while busily engaged searching amongst the branches, twigs, and buds of the trees and bushes for insect food, and its cheerful spring call-note of "Chicka-chicka, chee-chee-chee," seldom fails to attract attention.

During severe weather in winter, when the trees and hedgerows are covered with snow, it frequents the neighbourhood of farm-yards and houses, and may then be seen clinging round the eaves of stacks in search of food, or in company with sparrows, and other small birds, picking up seeds about the threshing-mill door—

> The birds flock silent and forlorn
> To barn-door step or granary eaves,
> And seek their scanty pittance there,
> For fields and hedgerows all are bare
> Of fruit or sheltering leaves.
>
> WALTER CHISHOLM.

It is a partial migrant, and is seen passing the lighthouses on the coasts of England and Scotland on migration in spring and autumn.[1]

---

[1] The Blue Titmouse was observed on migration in *Autumn* 1881.—At Farnes, 20th Oct. *Autumn* 1882.—Farnes, 8th Oct. *Spring* 1883.—Farnes, 8th March. *Autumn* 1883.—Farnes, 7th Oct. The above appear to be the only instances in the *Reports on the Migration of Birds*, as far as the lighthouses at the Farnes and Isle of May are concerned, where the Blue Titmouse has actually been identified. Doubtless many Blue Titmice are included among the "Titmice" which are pretty regularly reported as observed on migration at these two stations in spring and autumn.

THE BLUE TITMOUSE.

Its food consists principally of insects, such as caterpillars, grubs, and *aphides*, and it is doubtless a great benefactor to the gardener, although it has been accused of destroying buds in its search for insects.

It generally builds its nest in a hole in a wall, or in a tree, using grass, moss, hair, and feathers in its construction, and laying seven or eight eggs, which are white, spotted with light red. Mr. Weir of Boghead, Linlithgowshire, observed that a pair of Blue Titmice, on the 4th of July 1837, fed their young 475 times in the course of seventeen hours—from a quarter-past two o'clock in the morning until half-past eight in the evening, and that they appeared to feed them solely with caterpillars.[1]

---

[1] Macgillivray, *History of British Birds*, vol. ii. p. 438.

THE NUTHATCH.

NUTJOBBER, NUTHACK, WOODCRACKER.

Sitta cæsia.

Nuthatch piercing with strong bill.
SOUTHEY.

THE only instance on record of the occurrence of the Nuthatch in Berwickshire, is that mentioned by Mr. Gray in his *Birds of the West of Scotland*, on the authority of Dr. J. A. Smith. The specimen referred to was killed in a garden near Duns, in March 1856,[1] and sent to Dr. Smith to be exhibited at a meeting of the Royal Physical Society.

This bird is pretty common in the wooded parts of central and southern England, where it is found at every season of the year, but it is rare in the northern counties, and has been very seldom observed in Scotland.

It is partial to districts where there are old forest trees, the trunks and branches of which it climbs in search of insects, like the tree-creeper to which it bears an affinity. It runs up and down a tree with equal facility, and when it descends it keeps its head towards the ground. It does not use its tail as a support in climbing. The food of the Nuthatch consists of various insects, as well as hazel nuts, acorns, beech-mast, and other hard seeds. It fixes nuts in

---

[1] *Birds of the West of Scotland*, pp. 199, 200.

a chink of the bark of a tree, and hammers at them with the point of its bill until the shell is broken.

It builds its nest in a hole in a tree, the entrance, if too large, being plastered up with clay until it is of just sufficient size to admit the bird. The eggs, from five to seven in number, are white, spotted with reddish brown.

THE WREN.

KITTY WREN, COMMON WREN.

Troglodytes parvulus.

𝕮𝖍𝖊 𝕶𝖆𝖎𝖙𝖙𝖎𝖊-𝖜𝖆-𝖗𝖆𝖎𝖓, 𝕮𝖍𝖊 𝕶𝖆𝖎𝖙𝖙𝖎𝖊.

Beside the Redbreast's note, one other strain—
One summer strain, in wintry days is heard.
Amid the leafless thorn the merry Wren,
When icicles hang dripping from the roof,
Pipes her perennial lay; even when the flakes
Broad as her pinions fall, she lightly flies
Athwart the shower, and sings upon the wing.

GRAHAME.

THIS well-known and cheery little bird is almost as great a favourite with us as the Robin Redbreast, with which it is associated in several popular rhymes and superstitions, and which it somewhat resembles in its familiar and confiding habits. Although the Wren remains with us all the year round, yet it is a partial migrant; numbers being seen, while on migration, at the lighthouses on the Isle of May, and at the Farne Islands.[1] It suffered severely, and was greatly reduced in numbers, during the terrible winters of 1878-79, 1879-80, and 1880-81, many being found dead after the

---

[1] The Wren was observed on migration in *Autumn* 1880.—At Farne Islands, 23rd Nov. *Spring* 1881.—At Isle of May, 19th March. *Autumn* 1881.—Isle of May, 6th and 8th Oct.; Farnes, 4th Oct., many. *Autumn* 1882.—Isle of May, 10th Oct. and 12th Nov.; Farnes, 7th to 13th Oct. *Spring* 1883.—Isle of May, 9th April. *Autumn* 1883.—Isle of May, 22nd and 23rd Sep. and 7th Nov. *Spring* 1884.—Farnes, through March to 3rd April. *Autumn* 1884.—Isle of May, 17th Sep., 9th to 25th Oct. *Autumn* 1885.—Farnes, 23rd Oct. *Spring* 1886.—Farnes, 14th Feb. to 30th March.—*Reports on Migration of Birds*, 1879-86.

snow melted, in holes of walls and other hiding-places, into which they had apparently crept for shelter from the cold. Mr. Hardy, Oldcambus, writing in 1879 upon the effects of the previous winter, mentions that in his neighbourhood the destruction amongst Wrens was very great, and that they had almost disappeared from the woods and deans.[1] After the severe snow-storm of 1878-79 had passed away, three of these birds were found by me in holes of the policy-ground wall at Paxton, where they had sought shelter from the piercing cold, and been frozen to death.

In summer it frequents wooded deans, the margins of plantations, and the sides of the Whitadder, Leader, Eye, and other streams, where it creeps about amongst the roots of the alders, wild roses, and the like. It is also to be found at that season in places overgrown with brambles, whins, sloes, and other shrubs, where it is continually moving about, and

> Frae den to den,
> Gaes jinking through the thorn.
> TANNAHILL.

Towards the end of autumn it approaches our dwelling-houses and gardens, and may be often seen about outhouses and farm buildings, where, with cocked tail and bright eye, it makes itself at home, and when alarmed suddenly disappears in some hiding-place. Mr. Hardy says that it used to frequent the holes and caverns of the greywacke rocks on the sea-coast, until it was killed by the late severe winters.[2] The song of the Wren, like that of the Robin, is continued during the greater part of the year, and is remarkably loud for the size of the bird. Mr. James Smail mentions that it is often the first bird to commence singing in the morning on Leader

---

[1] *Hist. Ber. Nat. Club*, vol. ix. p. 125.
[2] I saw a Wren amongst the rocks near Fast Castle, on the evening of 30th June 1887, when I was watching the salmon-fishers drawing their nets in the sea there.

Water.[1] I have noticed that it holds its tail down when it sings on the branch of a tree.

The nest is comparatively large for the size of the bird, and is covered with a dome, having a small hole at the side for entrance. It is usually built of green moss, lined with hair and feathers, but the materials for the outside depend upon the situation in which it is found, which varies greatly.

> Among the dwellings framed by birds,
> In field or forest, with nice care,
> Is none that with the little Wren's
> In snugness may compare.
> WORDSWORTH.

It is sometimes placed against the trunk of a tree, in a mossy bank, or amongst the roots of alders or other bushes by the side of a stream; and I have seen it in the thatch of an old shed. The Wren occasionally chooses very odd situations for her home. A remarkable case of this kind was lately brought under my notice by Mr. Peter Cowe, Lochton, near Birgham, who informed me that, in the spring of 1869, a gamekeeper in his neighbourhood shot a carrion Crow, which he nailed up against a tree, the nail being driven through the head, and the Crow left hanging suspended from the nail, with its breast in contact with the bark. In the summer of 1870, a Wren's nest, with nine young ones, was discovered inside the skeleton of the Crow, the entrance being next the tree; and as the back, wings, and tail feathers of the Crow remained upon the skeleton, it could not be seen from the outside. Mr. Cowe has the skeleton of the Crow, with the nest in it, preserved in his collection of Berwickshire birds. It is well known that more nests of this species are built than are used for the reception of the eggs and young, a peculiarity regarding its breeding

---

[1] *Hist. Ber. Nat. Club*, vol. viii. p. 100.

habits which is alluded to by Darwin in his *Origin of Species*, as well as by Macgillivray, Yarrell, and Mr. Seebohm, in their Histories of British Birds. The last mentioned author says these "are widely known as *cock-nests*. Most country people, and not a few scientific naturalists, assert that they are either made for the male bird's reception, or that they are for the purpose of sheltering the birds during the inclement winter season. The explanation of this singular habit is still unknown, although many ingenious theories have been offered." It would appear, however, that they are sometimes used by Wrens for shelter at night, for Mr. Compton-Lundie of Spital, in the parish of Hutton, has informed me that, some years ago, when he was engaged one winter night in catching Sparrows with a net in the ivy on the wall of his house, he observed a bird fly out of a large Wren's nest in the ivy, and on putting his hand over the mouth of the nest, and taking it into a room in the house, no fewer than fifteen live Wrens were found in it. Macgillivray gives several instances where these birds have been found by his correspondents, sheltering in "cock-nests" at night, during severe weather in winter; and, from the letters which he quotes, it appears that these are not lined with feathers, like those in which the eggs are laid and hatched. The eggs, which are usually six or eight in number, are white, with small light red spots. The food of the Wren consists mostly of insects, but in autumn it will also eat fruit, and in winter crumbs and other refuse from the table, like the Robin. It was a popular belief at one time in some parts of Scotland that the Wren was the wife of the Robin, and an allegorical song in Herd's Collection refers to this—

> The Wren she lyes in care's bed,
> In care's bed, in care's bed;
> The Wren she lyes in care's bed,
> In meikle dule and pyne, O.

When in cam Robin Redbreist,
　　Redbreist, Redbreist ;
When in cam Robin Redbreist,
　　Wi' succar-saps and wine, O.

" Now, maiden, will ye taste o' this,
　　Taste o' this, taste o' this ;
Now, maiden, will ye taste o' this ?
　　'Tis succar-saps and wine, O."

" Na, ne'er a drap, Robin,
　　Robin, Robin ;
Na, ne'er a drap, Robin,
　　Though it were ne'er so fine, O."

" And where 's the ring that I gied ye,
　　That I gied ye, that I gied ye ;
And where 's the ring that I gied ye,
　　Ye little cutty quean, O."

" I gied it till an Ox-ee,
　　An Ox-ee, an Ox-ee,
I gied it till an Ox-ee,
　　A true sweetheart o' mine, O."

The Robin and the Wren, however, do not seem to have been quite free from occasional matrimonial difficulties, as the following quatrain attests—

　　The Robin Redbreast and the Wran,
　　Coost out about the parritch pan,
　　And ere the Robin got a spune,
　　The Wran she had the parritch dune.[1]

The Wren appears to have enjoyed some of the reverence[2] paid to the Robin by boys with regard to the harrying of its nest, and in some country places in Scotland they repeat the following malediction—

　　Malisons, malisons mair than ten,
　　That harry the Ladye o' Heaven's hen !

for such is the name given to our favourite by boys, even

---

[1] See *Popular Rhymes of Scotland*, by R. Chambers.

[2] The Wren had a sacred character amongst our Celtic ancestors.—Henderson's *Folk Lore of the Northern Counties*, p. 125.

THE WREN.

when engaged in the unhallowed sport of bird-nesting. It is also included in the following list of birds, whose nest it is deemed unlucky to molest :—

> The Laverock and the Lintie,
> The Robin and the Wren,
> If ye harry their nests,
> Ye'll never thrive again.[1]

---

[1] See *Popular Rhymes of Scotland*, by R. Chambers, New Edition, p. 187, 1888. For further particulars regarding the popular rhymes and superstitions connected with the Wren, see Macgillivray's *British Birds*, vol. iii. p. 19; Yarrell's *British Birds*, 4th Edition, vol. i. pp. 465, 466; *Folk Lore of the Northern Counties*, pp. 123, 124, 125; *Folk Lore of British Birds*, pp. 35-43; *Faune Populaire de la France*, Eugène Rolland, t. ii., "Les Oiseaux sauvages," pp. 288-301.

THE PIED WAGTAIL.

WATER WAGTAIL, WHITE WAGTAIL, BLACK AND WHITE WAGTAIL, WINTER WAGTAIL, PEGGY WASH-DISH, DISH-WASHER.

Motacilla lugubris.

𝕿𝖍𝖊 𝖂𝖆𝖙𝖊𝖗 𝖂𝖆𝖌𝖙𝖆𝖎𝖑, 𝕿𝖍𝖊 𝖂𝖎𝖑𝖑𝖞 𝖂𝖆𝖙𝖊𝖗 𝖂𝖆𝖌𝖙𝖆𝖎𝖑, 𝕿𝖍𝖊 𝖘𝖊𝖊𝖉-𝖇𝖎𝖗𝖉.[1]

> *What art thou made of?—air or light or dew?*
> *I have no time to tell you, if I knew.*
> *My tail—ask that—perhaps may solve the matter;*
> *I've missed three flies already by this clatter.*
>   MONTGOMERY, *Birds.*
>
> *Spare my grey beard, you Wagtail!*
>   SHAKESPEARE, *King Lear.*

THIS beautiful and elegant bird may be considered a regular migrant in Berwickshire, generally arriving in considerable numbers on the sea-coast near Cockburnspath and Oldcambus about the middle of March, and a little later in the inland districts of the county.[2] It leaves us in large flocks for the south in September and October, and is seen while on migration passing the lighthouses on the coasts of England and Scotland, including those at the Farne Islands

---

[1] It is known in some districts of Berwickshire as the "Seed-Bird," on account of its appearing at the time of seed-sowing in spring. Jamieson, in his *Scot. Dict.*, gives the meaning of "*Seed foullie*" as "the Wagtail, *mot. alba*"= *mot. lugubris.*

[2] Mr. Hardy's observations, extending from 1873 to 1882, show that its earliest spring arrival during that period was on 7th Feb. 1880, and the latest 3rd April 1876.

and the Isle of May, in spring and autumn;[1] but in mild seasons a few remain with us in sheltered places throughout the winter.

The Pied Wagtail is common throughout the county in summer, and is usually found in the vicinity of streams, pools, and marshy meadows, where it is always in motion, running after insects on the ground, and sometimes wading into the shallow water after them. It may be often seen flitting from place to place with an undulating flight, all the while uttering its twittering notes, and when not on the wing constantly vibrating its tail with a graceful motion. It is frequently noticed on the newly ploughed land in spring, when—

> at the ridge end stands the well-filled sack,
> And hive inverted, while the sower steps,
> With loaded sheet, along the furrowed ridge.
> GRAHAME, *British Georgics*.

Its nest, which is built of dead grass and fibrous roots, lined with hair, is found in a variety of situations, such as a hole in an old wall or bridge, the bank of a stream, and the like. A favourite spot for it in Paxton policy-ground is the wall of an old salmon-fishing bothy near the boat-house at the side of the Tweed.[2] I am indebted to Captain Logan Home of Broomhouse for the following interesting account of a very singular position

---

[1] The Pied Wagtail was observed on migration in *Spring* 1882.--At Farnes, 29th March. *Autumn* 1882.—Isle of May, 20th Sep., 31st Oct., and 7th Nov. *Spring* 1883.—Isle of May, 2nd and 30th March, 3rd and 25th April; Farnes, 2nd April, 4th and 14th May. *Autumn* 1883.—Isle of May, 11th Sep. to 10th Oct. *Spring* 1884.—Isle of May, 14th Feb., 9th and 22nd March; Farnes, 8th March. *Autumn* 1884.—Isle of May, 8th and 9th Sep. to 9th Oct. *Spring* 1885.—Isle of May, 24th Feb. to 23rd May; Farnes, 11th March to 9th May. *Autumn* 1885.—Isle of May, 20th to 29th Sep.—Extracts from *Reports on Migration of Birds*, 1879-1885.

[2] Mr. Hardy mentions nesting-places of the Pied Wagtail observed at the middle part of Edmond's Dean; at Heriot Water above Stockbridge; and on the Water Eye at Butterdean Mill; also at Renton and Horsley.—*Hist. Ber. Nat. Club*, vol. ix. p. 553.

which a pair chose for their nest. Writing from Cheltenham, on the 14th of June 1886, he says:—" I remember a pair of Pied Wagtails selecting a curious place for their nest at Broomhouse. Close to the hall door are two shells which my father brought from the *Morea Castle*, into which they had been fired. One is empty, the charge having been drawn, and in this empty shell the Wagtails built their nest, and brought up their family, in spite of people going in and out of the hall door frequently, and looking into the shell."

The eggs of the Pied Wagtail, which are four or five in number, are greyish white, finely spotted over with ash-colour. Mr. Hardy mentions that a White Wagtail (*Motacilla alba* of Linnæus) was seen " on stones in a burn" in the neighbourhood of Oldcambus on the 3rd of May 1881,[1] and Mr. George Pow, Dunbar, mentions that, in the spring of 1886, he observed a White Wagtail near the Pease Glen.[2] No specimen of the White Wagtail (*M. alba*) has, however, as yet (March 1888) been obtained in Berwickshire for identification.

---

[1] *Hist. Ber. Nat. Club*, vol. ix. p. 553. [2] *Ibid.* vol. xi. p. 544.

THE GREY WAGTAIL.

WINTER WAGTAIL, YELLOW WAGTAIL.

Motacilla sulphurea.

The Yellow Wagtail.

From mossy marsh and moorland scene,
Swift flows the Eye the hills between ;
By Quixwood old, and Butterdean,
It passes murmuring sweet.

DR. HENDERSON.

THE favourite haunts of the Grey Wagtail during the spring and summer months are the sides of our beautiful streams and burns, such as the Whitadder, Blackadder, Leader, Eye, Dye, Ale, Pease Burn, Dowlaw Burn, Blyth Water, and others, especially where their courses are rocky and gravelly. There it may be seen flitting from stone to stone, displaying its graceful form, or taking short hurried runs along the gravel in pursuit of the insects upon which it feeds. It is not very numerous, but a pair may be generally observed in the course of an hour's ramble along any of our river or burn sides, especially in wild and unfrequented districts, which the bird seems to prefer. A pair frequent the rocky burn which flows through the Old Heronry Wood at Paxton every year.

This species is a partial migrant in Berwickshire, most of those which breed here departing southwards in autumn, and only a few remaining with us during mild winters.

It returns in March and April,[1] and, shortly afterwards, commences to build its nest, which, although usually placed in an old wall, or the bank of a stream, is sometimes found at a little distance from water. As an instance of this I may mention that, in June 1883, Mr. Clapham, Broomhouse, near Duns, showed me a nest of this bird in a bed of sweet violets growing in a cold frame in the garden there. The nest is usually constructed of dry grass, roots, and moss, lined with hair, wool, or feathers; and the eggs, which are four or five in number, are greyish white, spotted all over with greyish brown.

Mr. George Bolam has informed me that he saw a Yellow Wagtail (*Motacilla raii*) at the side of the Whitadder below Clarabad Mill, on the 5th of May 1882. It was feeding on the grass haughland amongst some sheep, and was very shy and restless. No specimen of this Wagtail has, however, as yet (March 1888) been obtained in Berwickshire for identification. It appears to have been frequently observed about Dunbar in spring of late years.[2] The Yellow Wagtail (*M. raii*) may be easily distinguished from the Grey Wagtail (*M. sulphurea*) by the former bird being much shorter than the latter. The average length of the Grey Wagtail is $7\frac{3}{4}$ inches, while the length of the Yellow Wagtail is $6\frac{1}{2}$ inches.

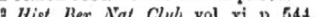

[1] Mr. Hardy, Oldcambus, records its arrival there 21st March 1874.—*Hist. Ber. Nat. Club*, vol. vii. p. 278. 3rd April 1876.—*Ibid.* vol. viii. p. 152. 4th April 1879.—*Ibid.* vol. ix. p. 129. 1st March 1882.—*Ibid.* vol. x. p. 556. Dr. Stuart, Chirnside, mentions seeing one at Allanton Bridge, on the Whitadder. 24th March 1884.—*Ibid.* vol. x. p. 576.

[2] *Hist. Ber. Nat. Club*, vol. xi. p. 544.

THE MEADOW PIPIT.

TITLARK, TITLING, MEADOW TITLING, MOSS CHEEPER,[1]
GREY CHEEPER, MEADOW LARK, PIPIT LARK,
MOOR PIPIT, LING-BIRD.

Anthus pratensis.

𝔗𝔥𝔢 𝔊𝔯𝔢𝔶 ℭ𝔥𝔢𝔢𝔭𝔢𝔯,[2] 𝔗𝔥𝔢 𝔗𝔦𝔱𝔩𝔦𝔫.

The titlene follouit the goilk, and gart hyr sing—guk, guk.
LINDSAY, *Complaynt of Scotland.*

THE Meadow Pipit is common on all our moors and upland pastures during summer, when it may be frequently seen rising from the ground, with its feeble cry of "Peep-peep-peep," and after flying a short distance with an undulating flight, again alighting amongst the heather or grass, or on some moorland stone or dyke. The Titling is associated in our minds with the wild uncultivated districts of the Lammermuirs—

> The round bare hill,
> The Law which ne'er has felt the plough,
> The pastoral slope and wimpling rill,
> The rushy bog and ferny knowe.[3]

It has given its name to the "Titlin Cairn," situate on the march of the Parishes of Longformacus and Lauder, between Hunt Law (1625 ft.) and Wedder Lairs (1593 ft.), two of the

---

[1] See Jamieson's *Scottish Dictionary.*
[2] Pronounced "Sheeper" in Berwickshire.
[3] Dr. Henderson's *MS. Poems.*

heights of the Lammermuir Hills. My friend Mr. Hardy has sent me the following lines on the Lady of Gamelshiels,[1] in which reference is made to its desolate haunts :—

> That scene is altered now and gone,
>   There desolation dwells;
> But still the stone on the moorland lone
>   The mournful story tells.
> There frequent now, on the wilds forlorn,
>   Wakes the Pipit's eerie trill;
> And the Curlew's call, at eve and morn,
>   Is clamorous and shrill.
> Oh woe! oh woe! to Gamelshiels Tower,
>   Oh woe! both night and day;
> For the fairest flower in the forest bower
>   Is vilely snatched away.

The Meadow Pipit is almost wholly migratory in Berwickshire, generally arriving on the sea-coast in the neighbourhood of Cockburnspath and Oldcambus in March, and spreading over the higher grounds in April. It may be observed frequenting bean stubble and turnip fields in the lower parts of the county in September and October, and shortly afterwards it leaves us for a milder climate,[2] but a few may be sometimes found throughout the winter in sheltered localities. During the severe snow-storm of December 1874, I observed several frequenting the marshy ground near the mouth of the Whitadder; and one day about the middle of the same month in 1878, when curling at Foulden, I saw one enter the house where the curling-stones were kept, for shelter. The poor bird soon became unable to fly from the biting cold, and was found dead when we took our curling-stones into the house in the evening.

---

[1] Gamelshiels is in the Lammermuirs, near Millknowe. The Lady was killed by a wolf. See *Hist. Ber. Nat. Club*, vol. iv. p. 291.

[2] Mr. Seebohm says that in autumn vast numbers pass along our eastern and southern coast, evidently on their migration southwards.—Seebohm, *Hist. Brit. Birds*, vol. ii. p. 226.

THE MEADOW PIPIT.

This species is observed on migration at the lighthouses on the coasts of England and Scotland in spring and autumn, and of late years it has been identified at the Isle of May and the Farne Islands.[1]

Its song is soft and musical, though not loud, and is generally uttered as the bird descends, with outspread wings and tail, from a little height to which it rises in the air before it commences to sing.

The nest, which is placed on the ground amongst grass or rough herbage, is composed of dry bents lined with hair, and the eggs, four or five in number, have the ground colour generally hidden with a close mottling of dark brown. The Cuckoo often deposits its egg in the nest of this species, and in the uplands of Berwickshire it is generally followed by several Meadow Pipits.

---

[1] It was observed on migration in *Spring* 1884.—At Isle of May, on 13th Feb. and 1st to 23rd April; at Farnes, 2nd April. *Autumn* 1884.—Isle of May, 16th Sep. to middle of October; Farnes, 14th Dec., large flocks. *Spring* 1885.—Isle of May, 10th to 18th March.—*Reports on the Migration of Birds*, 1879-85.

THE TREE PIPIT.

MEADOW LARK, TREE LARK, GRASSHOPPER LARK, PIPIT LARK,
SHORT-HEELED FIELD LARK, FIELD TITLING,
LESSER CRESTED LARK.

Anthus trivialis.

The Woodlark.[1]

> There is music uninformed by art
> In those wild notes, which, with a merry heart,
> The birds in unfrequented shades express.
>
> DRYDEN.

THIS interesting summer visitor generally arrives in Berwickshire from the south about the first week in May, and leaves us in September and October.

Shortly after its arrival in spring, it may be found thinly dispersed in pairs throughout the various wooded districts of the county, being partial to the neighbourhood of tall trees in deans, and the sides of streams. In this respect it differs from the Meadow Pipit, which inhabits open meadows and moors. The high elm and ash trees by the side of the Tweed near Finchy Shiel in the vicinity of Paxton, are a favourite resort of this bird. Mr. Hardy mentions that it frequents similar trees at Cockburnspath; also that it is seen in Penmanshiel Wood, Lamington Dean, Pease Dean, Dowlaw Dean, Aikieside

---

[1] Mr. Andrew Kelly mentions that the Tree Pipit is often mistaken for the true Woodlark (*Alauda arborea*) in Lauderdale, and is called the *Woodlark* there. —*Hist. Ber. Nat. Club*, vol. viii. p. 144.

Wood, and by the side of the Lambden Burn[1] near Antonshill.[2] It is recorded by Mr. Kelly as apparently increasing in Lauderdale.[3] Mr. John Thomson informs me that it nests at Cowdenknowes. I observed it amongst the old trees at Dryburgh Abbey, on the occasion of the visit of the Berwickshire Naturalists' Club to that beautiful and romantic spot on the 30th of June 1886, and a few days later—on the 6th of July—I noticed it in the Retreat Woods.

The song of the Tree Pipit is varied and sweet, and it often sings when in the air, as well as when sitting on the top of a tree. My attention was attracted one morning in June 1873, to the peculiar manner in which the bird sings when on the wing, for it flew up a short distance into the air from the top of a tree near my house, on which it was sitting, and then extended its wings backwards like a pigeon, and came slowly down to its perch, singing all the time until it alighted. Its food consists of insects and small seeds.

The nest, which is placed on the ground, is usually found amongst rough grass, and often near the root of a tree. It is built of moss, dry grass, and roots, lined with hair. The eggs, which are from four to six in number, vary much in colour, but they are generally greyish white, very closely spotted over with reddish brown.

The Tree Pipit is very like the Meadow Pipit in appearance, but a distinguishing mark of the former, when in the hand, is that the claw of the hind toe is slightly arched, and shorter than the toe itself, while in the latter the hind claw is as long as the toe.

---

[1] There is a popular rhyme in Berwickshire to the effect that
"The hooks and crooks o' Lambden Burn
Fill the bowie and fill the kirn."
DR. HENDERSON'S *Popular Rhymes of Berwickshire*.

[2] *Hist. Ber. Nat. Club*, vol. vii. p. 280; vol. ix. pp. 10, 554; vol. x. p. 251.
[3] *Ibid.* vol. viii. p. 144.

THE ROCK PIPIT.

ROCK LARK, SEA LARK, DUSKY LARK, SHORE PIPIT, SEA TITLING.

Anthus obscurus.

The Grey Sheeper.

Down to the sea—where plashing laves
The long sea-tangle amidst the waves;
Over yon bank where the wood-vetch clings
In its lovely bloom, and the Pipit sings.

DR. HENDERSON.

THIS species, which is considerably larger than the Meadow Pipit, frequents rocks by the sea-side, and is a permanent resident in Berwickshire; being found thinly scattered round our rocky coast during winter, and more plentifully in spring, summer, and autumn. Those which remain with us throughout the winter receive additions to their numbers in spring and autumn, in the shape of small migratory flocks; as has been observed by Mr. Hardy at Oldcambus,[2] where, on the 25th of April 1879, he saw a flock of about thirty alight on the bank behind St. Helen's Kirk, being apparently on their passage to the north.[3] It has been noticed on migration at the lighthouse on the Isle of May, in spring and autumn.[4]

---

[1] Called this on the coast about Greenheugh and Linkholm shore.—*Mr. Hardy's MS.*

[2] *Hist. Ber. Nat. Club*, vol. ix. pp. 130, 392; vol. x. p. 564.

[3] Mr. Hardy's MS. notes.

[4] The Rock Pipit was observed on migration at the Isle of May:—*Spring* 1883.—19th March. *Spring* 1884.—During March and April. *Autumn* 1884.- 13th Sep. to 4th Nov. *Spring* 1885.—4th April to 2nd May. *Autumn* 1885.- 19th Sep.- *Reports on the Migration of Birds*, 1879-86.

THE ROCK PIPIT.

In its manner of flight and song the Rock Pipit resembles the Meadow Pipit, but its usual haunts are very different, for it is seen only on the sea-coast and the immediate vicinity. Mr. Hardy remarks that there are always swarms of minute flies of the genus *Limosina* bred in the decaying sea-weeds, and that these constitute part of its food, and are picked up with alacrity. He adds that in spring and winter these Pipits resort to the sheep-folds and recently ploughed fields at some distance inland, in quest of insect food.[1]

The nest, which is composed of dry grass, is placed upon the ground, or upon some of the ledges of rock facing the sea, under the shelter of a tuft of grass. A nest was found amongst some whins in a turnip field at Oldcambus on the 23rd of July 1868. The eggs, which are four or five in number, are rather larger than those of the Meadow Pipit, or the Tree Pipit; and the ground colour, which is white, is usually thickly mottled over with greyish brown or olive. The hind claw is slightly longer than the hind toe.

---

[1] *Hist. Ber. Nat. Club*, vol. ix. p. 392.

THE GOLDEN ORIOLE.

Oriolus galbula.

The Blackbird and the Thrush,
The Golden Oriole, shall flit around,
And waken with a mellow gust of sound
The forest solemn hush.[1]

Rossignols, Loriots, Fauvettes,
Merles, Bouvreuils, Linots, Pinsons,
Cédant au pouvoir de mes sons.
Tous, jusqu'aux folles allouettes,
Venaient, pour prix de leur chansons.
De mon pain bequeter les miettes.

BERANGER, *Chansons.*

THE Rev. Andrew Baird, who drew up the report on the united parishes of Cockburnspath and Oldcambus for the *New Statistical Account of Scotland* in December 1834, says:—" And many a bird of fair and foreign plumage is

---

[1] The Rev. Alexander B. Grosart, in his *Poems and Literary Prose of Alexander Wilson, the American Ornithologist*, gives an interesting engraving of Wilson's tomb, which bears the following inscription:—"This monument covers the remains of Alexander Wilson, author of *American Ornithology*. He was born in Renfrewshire, Scotland, on 6th July 1766, migrated to the United States in the year 1794, and died at Philadelphia of the dysentery on the 23rd August 1813, aged 47." Mr. Grosart adds that a friend of his thus told the cause of Wilson's illness:—" While he was sitting in the house of one of his friends enjoying the pleasures of conversation, he chanced to see a bird of a rare species, for one of which he had long been in search. With his usual enthusiasm, he ran out, followed it, swam across a river over which it had flown, fired at, killed, and obtained the object of his eager pursuit; but caught a cold which, bringing on dysentery, ended in his death." The above lines are from a verse rendering of Wilson's dying wish to be laid in some rural spot where the birds might sing over his grave.—*Poems and Literary Prose of Alexander Wilson, the American Ornithologist*, by the Rev. Alexander B. Grosart, 2 vols. Paisley, 1876.

THE GOLDEN ORIOLE.

occasionally seen to halt for a day or two among the woody retreats of the parish. Of these occasional visitants we may notice in particular the Bohemian Chatterer, the Hoopoe, and the Golden Oriole."[1] There is no other record of the appearance of this species in Berwickshire, and it is so very rarely seen north of the Tweed that Mr. Gray mentions only four instances of its occurrence in Scotland besides the above.[2]

I have seen this beautiful bird and heard its rich flute-like notes in the neighbourhood of Romarantin, Loir et Cher, France, where it is plentiful during summer. It frequented the tops of the tall oak and beech trees in the parks which surrounded the country houses of the landed proprietors, and was exceedingly shy and difficult to approach. On referring to my note-book, I find that, on the 4th of July 1870, I endeavoured to obtain a specimen in the park which surrounded the residence of M. Jullien, near La Ferté Ambault. The Orioles kept continually flying from the top of one clump of trees to that of another, always alighting amongst the leafy branches, and taking wing long before I could get within gunshot of them. All my endeavours to get within range, by following them, were fruitless. At last I concealed myself behind the trunk of a tall beech in the centre of one of the clumps, and my brother, who accompanied me, walked round to the opposite side of the park, and drove the birds in my direction. After waiting for some little time, I had the satisfaction of seeing a beautiful yellow male alight on the top of a tree near my place of concealment, and of securing it. The heat in the park was oppressive, being 95° in the shade, and the barrels of my gun became so hot when exposed to the

---

[1] *New Statistical Account of Scotland*, vol. ii., "Berwickshire," p. 299.
[2] *Birds of the West of Scotland*, pp. 80, 512.

rays of the sun that I could not hold them in my hand. The French give the following interpretation to the song :—

> Berlusiau, berlusiau,
> Qui mange la cerise,
> Et laisse le grimiau.[1]

The Golden Oriole is about the size of a Blackbird, and the plumage of the male, with the exception of the wings, which are black, is bright yellow. It is a spring and summer visitor to countries north of the Mediterranean, when a few stragglers reach the British Islands. Its nest has been found in England.

---

[1] Rolland, *Faune populaire de la France*, tome ii., " Les Oiseaux sauvages," p. 233.

Cockburnspath

THE GREAT GREY SHRIKE.

BUTCHER BIRD, MURDERING PIE.

Lanius excubitor.

The Butcher Bird.

Spare helpless innocence—
Troth, pleasant talk!
Yon Sparrow snaps more lives in a day,
Than in a twelvemonth I could take away.
MONTGOMERY, *Birds.*

L'oiseau bruyant, hardi, qu'on nomme Pie-Grièche,
Nous rappelle à, l'esprit femme à l'humeur revêche.[1]

THIS species is sometimes seen in the county during the autumn, winter, and spring months, but its visits are rare and irregular.

In the report on the parish of Eccles, in the *New Statistical Account of Scotland*, 1834, Dr. Robert D. Thomson, F.R.S., mentions that the parish is occasionally visited by the Greater Butcher Bird.[2] Mr. Wilson, late of Edington Mains, informs me that his brother shot a specimen of this bird on a thorn tree near the public road there in 1845. Mr. William Patterson tells me that he killed a Grey Shrike many years ago, at Grueldykes, near Duns, and also got two—a male and a female—a few years later, on the farm of Greenhead, in the neighbourhood of Reston. He says they were

---

[1] The French call a shrewish woman a Pie-grièche, or Shrike.
[2] *New Stat. Acc. of Scot.*, vol. ii., "Berwickshire," p. 53.

very shy and wary, and appeared to have a liking for perching on the topmost twigs of the hedges. Mr. Kelly records that Mr. Tilly, Lauder, succeeded in shooting a beautiful specimen of this unusual visitant, which was in company with a Magpie, late in October 1872, on the estate of Allanbank, near Lauder. He adds that Mr. Tilly could not understand what was the matter with all the little birds, as they were flying about in a strange manner and making a loud outcry; but on advancing he saw a Butcher Bird dash away, and, after a number of jerks and doublings among the trees, it fairly gave its persecutors the slip.[1] A male Butcher Bird was shot at Lochton in the parish of Eccles in the end of December 1875.[2] In November 1876, a Shrike of this kind was found dead in a garden at Duns, and another was seen at the same time, about a mile further south.[3] A male was shot near Swinton on the 5th of April 1878;[4] and Dr. Stuart mentions that while his son was driving to Greenburn on the 13th of February 1881, when near Auchencrow Mains, he observed a peculiar bird on one of the hedgerow trees, which sat until he satisfied himself that it was a Great Grey Shrike.[5] An example of this species was found dead at Fans, in the parish of Earlston, on the 20th of November 1881;[6] and another was killed at Gordon, on the 6th of February 1883.[7]

It feeds on small birds, mice, frogs, and large insects, and, after killing its prey, often fixes the body in a forked branch, or on a sharp thorn. From this habit it has derived the name of Butcher Bird. My friend, Mr. A. P. Hope of Sunwick, lately presented me with a fine specimen which he had shot at Fentonbarns, East-Lothian, about

---

[1] *Hist. Ber. Nat. Club*, vol. vii. p. 303.
[2] *Ibid.* vol. vii. p. 500.
[3] *Ibid.* vol. viii. p. 196.
[4] *Ibid.* vol. viii. p. 524.
[5] *Ibid.* vol. ix. p. 405.
[6] *Ibid.* vol. ix. p. 562.
[7] *Ibid.* vol. x. p. 572.

Christmas 1870, when snow was lying on the ground. His attention was attracted to the Shrike by seeing it flying with a bird in its claws, which proved to be a lark.

The Great Grey Shrike is used by Dutch falconers at Falconswaerd, to assist in catching wild Peregrine Falcons. It is tethered near the clap-net used by them for taking the Hawks, and it gives notice of the approach of a Hawk by its cries; when the falconer, being thus warned, displays a live pigeon, which has been kept concealed under sods of turf, and the moment the Hawk seizes the Pigeon, the net is drawn over them both.[1]

The size is about that of a Blackbird. The upper plumage is of a pearl-grey colour, wing coverts black, quill feathers of the wing black, with a white bar at the base, forming, when the wing is closed, two white spots. I may here draw attention to the fact that Mr. Seebohm mentions the occurrence of Pallas's Grey Shrike (*Lanius major*) in Forfarshire, in the winter of 1869, and adds that it appears to be a comparatively frequent visitor to our islands, though it has hitherto been overlooked, probably in consequence of its similarity to the subject of our article. He says:—" In the Great Grey Shrike (*Lanius excubitor*) the outer webs of the primaries are white at the base, and thus, when the wing is closed, a white bar is formed across the part formed by the primaries. In Pallas's Grey Shrike (*Lanius major*) exactly the same occurs, so that, so far as the primaries are concerned, there is no difference in the two species. When we come to look at the secondaries, we shall, however, find that in *Lanius excubitor* the bars of their outside webs are also white, so that two white patches or bars are formed on the wings, whilst in *L. major*, there is no white at the base of the outside webs of the secondaries,

---

[1] A most interesting account of this method of taking wild Falcons is given in *The Field* newspaper of 16th Feb. and 16th March 1878.

so that only one white bar or patch is formed on the wing."[1]

The Great Grey Shrike has not been ascertained to breed in Great Britain.

---

[1] *Proceedings of the Royal Physical Society*, Session 1881-82, vol. vii. pp. 223, 224.

THE RED-BACKED SHRIKE.

BUTCHER BIRD, MURDERING PIE, FRENCH MAGPIE, FLUSHER.

Lanius collurio.

*The Mayfly is torn by the Swallow, the Sparrow is speared by the Shrike,
And the whole little wood where I sit is a world of plunder and prey.*

TENNYSON.

THIS bird appears to have been very seldom observed in Berwickshire, although it is a well-known and regular summer visitor to various districts in England.

The Rev. John Duns, Torphichen, in a communication to the Royal Society of Edinburgh, mentions the occurrence of two specimens at Oxendean, near Duns Castle, in July 1859;[1] and Dr. Turnbull relates that Lord Binning saw a male on the farm of Byrewalls, near Gordon, in the autumn of 1865.[2]

The Red-Backed Shrike visits England about the end of April or beginning of May, where it breeds. It leaves in September for the south.

Its food consists of mice, small birds, and various insects.

An interesting account of the habits of this bird is given in *The Field* of the 28th of November 1885. The writer says :—" Having had a Red-Backed Butcher Bird for the last two and a half months, I have had ample opportunity of

---

[1] *Proceedings of the Royal Society of Edinburgh*, vol. iv. 1861-62, No. 57, p. 532.

[2] *The Birds of East-Lothian*, by W. P. Turnbull, 1867, p. 38.

observing the tactics of this bird with regard to its prey. Nature has given the Shrike weak feet and claws, but has compensated for this defect by giving it the art of hanging up its prey on a thorn and thus getting the required leverage to tear it up piecemeal. I have fixed up in the cage of my bird a thorn bough, sharpened into spikes, on which it hangs up everything too large to swallow at once. On a small bird or mouse being given, it commences operations by breaking up the head with its strong bill; and in the case of a mouse it will smash up the whole of the skull before it can get through the skin to eat it. It almost invariably flies up with its prey in its beak, and always spikes it through the neck, but on one occasion it persistently flew up with a Greenfinch in its foot. The spikes had become very blunt with bits of fur and skin, and the Greenfinch would fall every time the Shrike let go, thinking he had it nicely fixed up. It occasionally takes up a piece of meat to its perch, and tries to eat it while grasped in one foot. While doing so the heel is rested on the perch, and in this position it can do little more than pull the meat with its own foot up and down aimlessly. I have never seen it stand upon its prey and tear it like a Hawk. Feathers, fur, etc., it casts up in pellets, after the manner of Hawks and Owls. A friend of mine has in his collection a Red-Backed Shrike which was shot in the act of flying away with a young Partridge." Another writer in the same paper says :—" Having kept three species in confinement—viz., the Great Grey, Woodchat, and Red-Backed—I can assert that they hold their food in one foot, resting the elbow of that leg on the perch to steady themselves, and tearing it with their powerful beaks. They do not place the food on the perch and put their foot on it as a Hawk would do, and in flying with it they invariably carry it in their beak. Food is only hung up for future use."

THE WAXWING.

BOHEMIAN WAXWING, BOHEMIAN CHATTERER, SILK TAIL, WAXEN CHATTERER.

Ampelis garrulus.

*The glossy Finches chatter
Up and down, up and down.*

JEAN INGELOW.

THIS beautiful bird occasionally visits Berwickshire during winter and spring, at uncertain intervals. The earliest notice of its occurrence in the county is that by the Rev. Andrew Baird, in his report on the united parishes of Cockburnspath and Oldcambus, in the *New Statistical Account of Scotland*, written in 1834. In the following year the Rev. John Turnbull of Eyemouth, in his report on that parish in the same publication, mentions that it had been sometimes seen in the plantations round Netherbyres. In December 1835 one was caught near Coldstream, and kept in a cage for some time by the late Dr. Johnston of Berwick, who, in a letter to his friend, Mr. Joshua Alder, of Newcastle-on-Tyne, says:—"Tell Mr. A. Hancock that I have a fine Waxwing chirping away merrily."[1] Major-General Cockburn-Hood of Stainrigg has informed me that a specimen was shot on his estate about 1851. In April 1883, I saw in a collection of birds made by the

---

[1] I am indebted for this information to Mrs. Barwell-Carter, of The Anchorage, Berwick, daughter of the late Dr. Johnston, who possesses a very beautiful coloured drawing of the bird by the late Mrs. Johnston.

late Mr. Simpson, Lauder, an example of this species which had been obtained near Lauder-Barns about 1860. Mr. Hardy, Oldcambus, writing in 1872, says that several years previous to that date his brother saw two Waxwings near the post-road between Grantshouse and Penmanshiel; and he also mentions, on the authority of Mr. Wilson of Coldingham, that one was seen at Hallydown on the 11th of May 1872, and another killed at Coldingham shortly afterwards.[1] Mr. John Ferguson, Duns, records that one was shot on Duns Castle estate in the winter of 1873.[2]

Great numbers of Waxwings appeared in various parts of the country in the winters of 1830-31, 1834-35, 1849-50, and 1866-67.

The food is principally berries and insects, and the bird is very voracious.

Nothing was known of its nidification until 1856, when it was discovered by Mr. John Wolley breeding in Lapland. It nests in the pine regions near the Arctic Circle.

The Waxwing, which is about the size of a Starling, derives its name from the shafts of the secondary feathers of the wing being tipped at the lower end with a substance which resembles scarlet sealing-wax.

---

[1] *Hist. Ber. Nat. Club*, vol. vi. p. 427.   [2] *Ibid.* vol. vii. p. 284.

THE SPOTTED FLYCATCHER.

Muscicapa grisola.

The Flycatcher, The Egypt.[1]

Summer birds pursuing gilded flies.
COWPER, *Winter Walk at Noon.*

THE Spotted Flycatcher is one of our latest summer migrants, and does not often make its appearance until the advent of mild weather, when the cold breath of winter has passed away for the season.

It generally comes from about the middle to the end of May, when the trees are assuming their summer dress; but it has been observed in the county as early as the end of April.[2] It leaves us for a warmer climate about the beginning of September, and appears to spend the winter months in Africa.

It is a modest, silent, unobtrusive bird, and during summer may frequently be seen perched on a paling or post, or the dead branch of a tree by the roadside, and at short intervals darting away after insects for a few yards at a time, and then returning to its perch again.

It frequents roadside trees, avenues, and gardens, and seems to be in no way particular about the position of its nest, as far as concealment is concerned; for it often builds on the boughs of wall fruit trees, close to walks, or in other

---

[1] This local name has been given to the bird on account of its note being like the word "Egypt."
[2] Dr. Stuart records that he saw several Spotted Flycatchers at Blanerne Bridge hawking for flies on 30th April 1883.—*Hist. Ber. Nat. Club*, vol. x. p. 571.

exposed situations. It is sometimes very whimsical in the selection of a nesting-place, and, as an instance of this peculiarity, I may mention that, a few years ago, a pair built in a tea-cup which had been left standing for some weeks on the outside of the window-sill of one of the garden offices at Milne Graden. During incubation the cup was several times removed from the sill by the gardener to show its contents to visitors, and yet, notwithstanding this disturbance, the birds succeeded in rearing a brood of young in the strange home which they had chosen. The nest is composed of moss, small twigs, and roots, and is lined with horse-hair, wool, and feathers. The eggs are four or five in number, and vary in ground colour from bluish white to pale greenish blue, blotched and speckled with reddish brown.

The Spotted Flycatcher feeds exclusively on insects. It has no note beyond a kind of faint chirp, like the word "Egypt," which it utters at intervals.

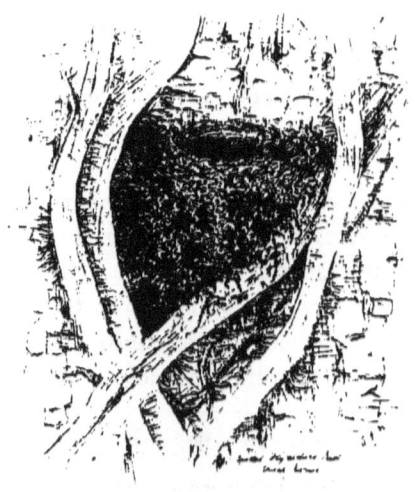

THE PIED FLYCATCHER.

Muscicapa atricapilla.

> *When May is in his prime, and youthful Spring*
> *Doth clothe the trees with leaves, and ground with flowers,*
> *And time of year reviveth everything,*
> *And lovely Nature smiles and nothing lowers.*
>
> THOMAS WATSON, 1581.

THIS pretty bird is not a regular spring visitor to the county, several years sometimes elapsing between its visits; and when it does make its appearance, which is generally about the middle of May, it seems to be only crossing the district on the way to its breeding quarters. There is no record of the nest having been found in Berwickshire, but it has been discovered in the adjoining county of Northumberland.[1]

The Pied Flycatcher appears to be a local bird in England, only frequenting certain districts, the lake country being a favourite resort.

Mr. Gray mentions that Dr. Smith, of Edinburgh, had informed him that a male bird of this species was shot by Mr. Stevenson in a garden near Duns, in the first week of June 1855, and that another male was seen in the same place, in June of the following year.[2]

In the second week of May 1872, I saw a male Pied Flycatcher in a plantation by the side of the Tweed at

---

[1] *The Birds of Northumberland and Durham*, by John Hancock, p. 79.
[2] *Birds of the West of Scotland*, p. 69.

Finchy, and one was noticed near the Avenue Bridge, in the grounds of Paxton House, on the 12th of May 1877. Mr. Small, Bird-stuffer, Edinburgh, has informed me that, in 1874, he preserved one which was shot near Coldingham.

In May 1884 this bird appeared in small numbers in various districts of the county. Three beautiful specimens were observed by me in the grounds of Paxton House in the second week of that month; and Mr. Andrew Balsillie, Dykegatehead, informs me that he saw a pair in the Pistol Plantation about the same time. In the beginning of May 1886 one was noticed in the Retreat woods by Mr. James Shiel, gamekeeper, Abbey St. Bathans.

The nuptial dress of the male on his arrival in spring is conspicuous, being black and white, and the bird consequently is very noticeable at that season. After the autumnal moult he assumes somewhat the same plumage as the female and the young of the year, which is of an obscure brown colour, and not attractive to the eye. In this state several, both old and young, have been observed in the months of September and October near the sea-coast at Berwick,[1] being evidently on their return journey to the south. Like the Spotted Flycatcher, it appears to spend the winter months in Africa.

The Pied Flycatcher during its spring visit seems to be fond of frequenting the edges of woods, avenues, trees overhanging small streams, and other spots where

> The hollow whispering breeze, the plaint of rills,
> That, purling down amid the twisted roots
> Which creep around, their dewy murmurs shake
> On the soothed ear.
> THOMSON, *Spring*.

---

[1] *Hist. Ber. Nat. Club*, vol. ix. p. 560, and vol. x. p. 386.

THE SWALLOW.

CHIMNEY SWALLOW, COMMON SWALLOW, RED-FRONTED SWALLOW, BARN SWALLOW.

Hirundo rustica.

The Swallow.

> *The Swallow knows her time,*
> *And on the vernal breezes wings her way*
> *O'er mountain, plain, and far-extending seas,*
> *From Afric's torrid sands to Britain's shore.*
>
> GRAHAME, *Birds of Scotland.*

OF all our migratory birds the Swallow is one of the greatest favourites, for it nestles under our roofs, and on its return in spring is ever welcome as the harbinger of summer. After its winter sojourn amongst the palm-trees of Africa [1] or Palestine, this clime-changing bird, hearing the voice of spring, remembers that the time for revisiting its home in the north has arrived; and, setting forth on its aërial journey, flies swiftly along the coasts of Northern Africa,[2] and over sunny Spain and France; or, crossing the Mediterranean to

---

[1] Mr. Hardy writes that an old Berwickshire story about Swallows is told in the following form:—"A man, to ascertain where they went to in winter, caught one, and placed round its neck a piece of parchment inscribed 'Kimmerghame Mill.' It returned in the following spring with the reply, 'River Nile, Egypt.'"

[2] The Swallows disappear suddenly, they are followed across the Channel, watched at Gibraltar, reported coasting along the shores of North Africa, and then duly cabled as catching flies and twittering at Cairo or Jerusalem.—*The Migration of Birds,* in *Edinburgh Review,* Jan. 1885. [The return journey in spring will be along the same route.]

the land of the myrtle and olive, it passes over the Alps by well-known routes,[1] and speedily reaches our shores.

> The Swallow dreams on Afric's shore
>   Of Scotia's summer pride,
> And plumes her wing and knows the hour
>   To hasten to Tweedside.
>
> The cliff or scaur she kens afar,
>   And towering ruins grey,
> Where she was nursed in Dryburgh's bower
>   The lap of flowery May.
>
> JOHN YOUNGER.

Its arrival in Berwickshire generally takes place about the 26th of April, the earliest record for the last eighty-five years being on the 11th of April 1803, and the latest on the 11th of May 1850.[2] It has usually paired when it comes, and shortly afterwards, with much joyous twittering, it proceeds to mend the old nest of the previous year, or to select a suitable site for a new one, which is often under the open roof of some barn, shed, or other building, and sometimes on the rafters beneath the eaves of a dwelling-house.[3]

The Rev. Gilbert White, in his delightful *Natural History of Selborne*, mentions that in his time the Swallow constantly built in chimneys; and this appears to have been its habit long ago in Berwickshire also, when the old wide "lums" were in use. Mr. Hardy relates that "in 1841 there was a room in a house in Whitsome village in which there had not been a fire for a long time, and when one was at last lighted the vent was found to be stopped.

---

[1] "It is well known to naturalists that Swallows cross the Alps by certain passes, as is also done by other species, regularly every year. I am informed by Mr. A. B. Herbert that some of the favourite Alpine passes for the annual migration of birds are the Albula and Bernina Passes into the Adda Valley and Lake Como, returning the same way in spring."—J. A. Harvie-Brown, in *Fourth Report on the Migration of Birds*, 1882, p. 70.

[2] See *Ornithological Calendar for Berwickshire*, in the Appendix, at the end of vol. ii. of this work.

[3] Several pairs have nested on the ends of the rafters under the eaves of my house at Paxton for the last seven or eight years.

The draught of warm air, however, loosened a quantity of rubbish in the chimney, among which were the remains of a nest of Swallows that had been hatched too late, and had been left behind by the parents to perish. In another house in the same village, the birds had frequented a chimney for more than fifty years.[1] In 1846 a pair built in the chimney of an outhouse attached to an inn in Ayton, a fire being lighted only on 'washing days,' when the boys of the village used to watch the exit and entry of the birds at the top of the chimney. At an old mill near that village a pair were accustomed to build regularly every year on a projecting piece of wood in the top of the chimney of the kiln, in the midst of smoke and no slight heat." Mr. John Pringle, joiner, Paxton House, has informed me that, about forty-five years ago, Swallows used to build in the chimney of his workshop in Paxton village, which is one of the old wide kind in use long ago, and where a fire was only occasionally lighted in winter.[2]

> The Swallow . . .
> Is well to chancels and to chimneys known,
> Though 'tis not thought she feeds on smoke alone.
> DRYDEN.

The saucer-like nest is formed of small pieces of moist earth collected by the bird at the sides of ponds or pools, and built, with the addition of short pieces of straw and stick, into the required shape. After the nest has been warmly lined with feathers, the eggs, which are from four to six in number, and are white, blotched and speckled

---

[1] Mr. Lees, Public School, Whitsome, informed me on 26th August 1887, that Mr. James Grieve, shoemaker in that village, who is eighty years old, remembers Swallows building in the chimneys of Whitsome village fifty or sixty years ago.

[2] Although the chimney still remains as it was long ago, no Swallows have built in it for forty years. It would appear that owing to the general improvement which has taken place in the construction of chimneys in the county, the Swallows have changed their former habits, for they do not now nest in chimneys.

with ashy-grey and brown, are laid ; and while the female is sitting the male frequently visits her, and cheers her with his song. When the young are hatched the parents are continually on the wing in search of food for their nestlings from early dawn [1] until it becomes dark, and may then be seen sweeping round houses and other buildings, skimming over grass fields where cattle are feeding, along the edges of woods, or over the surface of water, and returning to the nest every few minutes to feed the young. Two broods are reared in the season. The young of the first brood, after leaving the place where they were reared, collect together into flocks, and may be seen resting on telegraph wires and roosting on trees. They are believed to migrate southwards in the end of August and beginning of September, and are followed by the old birds and the young of the second brood towards the end of the latter month. Shortly before this time these may be seen assembling into great flocks, and sitting on the roofs of churches,[2] barns, and other high buildings, such as Marchmont House, and the County Buildings at Greenlaw, preparatory to their southern journey.

> The Swallows now disown
> The roofs they loved before ;
> Each, like his tuneful genius, flown
> To greet some happier shore.
>
> SHENSTONE.

After the main body has left, a few sometimes linger on a little later, and, in 1867, one was observed at The Hirsel as late as the 10th of November.[3]

---

[1] I have heard them twittering on the wing round my house at Paxton between one and two o'clock in the morning in July.

[2] The late Rev. William Stobbs, Gordon, records that the favourite rendezvous of the Swallows in that district, before setting out for the south, is the church roof at Gordon, or the steading at Greenknowe.

[3] The following anecdote of a Swallow was communicated to the Berwickshire Naturalists' Club by the late Earl of Home :—" On the 7th Nov. last 1867], long after the Swallows had taken their departure, a Swallow, probably

The Swallow appears to have a great antipathy to birds of prey, and may be often seen flying after hawks high in the air : it has also been observed swooping down at cats. It is said, in Berwickshire, that when Swallows fly very low, and sweep close over pools of water, rain may be expected. This habit of the bird appears to have been regarded as one of the prognostics of a storm as early as the days of Virgil, for we find him saying in his *Georgics*—

> Numquam imprudentibus imber
> Obfuit : aut illum surgentem vallibus imis
> Aëriæ fugere grues ; aut bucula, cœlum
> Suspiciens, patulis captavit naribus auras ;
> Aut arguta lacus circumvolitavit hirundo ;
> Et veterem in limo ranæ cecinere querelam.

When they fly high in the air it is believed that the weather will be fine.

> When Swallows fleet soar high and sport in air,
> He told us that the welkin would be clear
> GAY.

"There is," says Dr. Henderson, of Chirnside, " a singular superstition prevalent amongst boys of the county to the effect that if a Swallow fly through betwixt a person's arm and his body, the person will lose the power of the arm for ever. It is obvious that there can be little danger of such a thing ever taking place, yet I remember that, when I was a boy, I had a great dread of the Swallow when she was skimming past me on her swift pinions, and always

---

a young one left behind by the others, came into my library [at The Hirsel', and perched on the rod of a window curtain, and passed the night. When the window was opened in the morning it went out, and was observed flying about all day feeding on insects ; the frost had been sharp, thermometer 20 in the night. In the evening the Swallow returned, and took up its perch on the same spot. On the 9th it pursued exactly the same course, returning to its perch. On the 10th the weather changed and became milder, the bird went out as usual, but was not seen again, and probably set off to try and overtake its companions."
– *Hist. Ber. Nat. Club*, vol. v. p. 142.

took good care to keep my arms close to my body, in accordance with the advice in the following quatrain,—

> Quick, quick, the Swallow comes,
> Keep your arms close,
> For if she touch them wi' her wings,
> You their power will lose.

Mr. Hardy notes that, in some parts of the county, this bird is said "to drink a drap o' the deil's blood every morning." A curious superstition is mentioned by Mr. Lockie, Spottiswoode, who says:—"There is a belief in our district with reference to the killing of a Swallow, or the robbing of its nest, that the hair of a person who does so will not grow until the return of the Swallows in the following year, and that he will have frequent headaches in the interval." It is a common saying in the county that "one Swallow does not make a summer."[1] This appears to have had its origin in the changeable nature of the weather about the time of its arrival; a few mild days, which may have induced some of the first comers to visit a locality, being followed by a spell of cold ungenial weather, such as we often experience about the end of April and beginning of May, so aptly described by Dunbar,—

> For yistirday, I did declair
> How that the sasoun fast and fair
> Come in als fresche as pacock feddir,
> This day it stangis lyk ane eddir.

---

[1] Mr. A. H. Evans, Cambridge, remarks:—"This is an old saying of the Greeks, though they said 'spring' instead of summer."

THE MARTIN.

HOUSE SWALLOW, WINDOW SWALLOW, HOUSE MARTIN, MARTINET, MARTLET, WHITE-RUMPED SWALLOW.

Chelidon urbica.

𝕿𝖍𝖊 𝖂𝖎𝖓𝖉𝖔𝖜 𝕾𝖜𝖆𝖑𝖑𝖔𝖜, 𝕿𝖍𝖊 𝕾𝖊𝖆 𝕾𝖜𝖆𝖑𝖑𝖔𝖜.

> *This guest of Summer,*
> *The temple-haunting Martlet doth approve,*
> *By his loved mansionry, that the heaven's breath*
> *Smells wooingly here; no jutty frieze,*
> *Buttress, nor coign of vantage, but this bird*
> *Hath made his pendant bed, and procreant cradle:*
> *Where they most breed and haunt, I have observed,*
> *The air is delicate.*
> — SHAKESPEARE, *Macbeth.*

THE Martin or Window Swallow generally arrives in the county a little later in spring than the Chimney Swallow,[1] and, like the latter, is a much loved bird. Its pleasing appearance and manners, as well as its familiarity in building its nest in the corners of our windows, render it a general favourite; while its soft, cheerful twitter, heard in the dawn of early summer mornings, is associated in our minds with the most delightful period of the year—

> When May is in his prime, and youthful Spring
> Doth clothe the trees with leaves, and ground with flowers.
> — THOMAS WATSON, 1581.

The Martin differs from the Chimney Swallow in its nesting habits, and it may likewise be easily distinguished from the

---

[1] The earliest date of its arrival in Berwickshire for the last fifty years was on the 21st of April 1874, and the latest on the 14th of May 1877. See *Ornithological Calendar of Berwickshire*, in the Appendix, at the end of vol. ii. of this work.

latter bird when on wing, by its white rump and shorter tail. While the Chimney Swallow confines itself wholly to our buildings for the site of its nest, great numbers of our Martins continue to build in the situations which they would occupy long before there were any houses in Berwickshire, such as against the sides of the rocky caverns and precipitous cliffs of the sea-coast, from Swallow Craig, near Oldcambus, round by Fast Castle and St. Abb's Head, to the neighbourhood of Burnmouth. They also frequently make their nests against the precipices which overhang our rivers, the steep rock on the north bank of the Whitadder, at Edrington, being a favourite resort. Shortly after its arrival in spring, the bird either repairs an old nest or builds a new one, which, being composed of mud mixed with little bits of straw, and plastered against the corner of a window, or the side of a wall or rock, without any support, requires to be securely built to prevent it from falling. The bird wisely takes care to allow the lower layers of mud to harden before more are added—

> Whence drew the Martin his superior skill
> To knead and temper, mason-like, the slime
> Of street or stagnant pool, and build aloft
> Beneath the cornice brink or shady porch,
> His snug depending couch, on nothing hung?
>   Hurdis.

The nest is not open at the top like that of the Chimney Swallow, but has a small hole in the upper part of the side for the entrance and exit of the bird. The inside is lined with fine grass and feathers, and the eggs, which are four or five in number, are pure white. The building of the nest is generally completed in ten or twelve days, and incubation lasts thirteen days. Two broods are commonly reared in the season. The young of the second brood are sometimes found dead in the nest in autumn, the parents having either been unable to find a sufficient supply of insects to feed their offspring, owing to the lateness of the season, or having abandoned the young, the sudden occurrence of cold weather and consequent failure of food having forced them to migrate southwards.[1]

The young of the first brood congregate together, and may be seen sitting in flocks on the top of buildings [2] in August, when they are supposed to proceed to the south; the parents and the fledged young of the second brood following later in the season. The food of the Martin consists wholly of insects. Mr. Seebohm says that scarcely anything is known of the winter quarters of the Martin, which are probably somewhere in Central Africa.[3]

Swallow Craig, a steep precipice on the sea-coast near Oldcambus, Swallowheugh near Chirnside, and Swallowdean in the neighbourhood of Broomhouse, have apparently derived their names from being frequented by numbers of the Martin or Window Swallow.

---

[1] Darwin, in referring to the instincts of animals in his *Descent of Man*, says:—"But the most curious instance known to me of one instinct conquering another is the migratory instinct conquering the maternal instinct. The migratory instinct is so powerful that late in autumn Swallows and House Martins frequently desert their tender young, leaving them to perish miserably in their nests."— Vol. i., London 1871, pp. 83, 84.

[2] Mr. Hardy mentions that large numbers of old and young Window Swallows were observed assembled on the top of Coldingham Church on 10th August 1857. He also says that they have a gathering-place at Bleakheugh on the coast, near Gunsgreen, whence they set out on their journey southwards in autumn.— *MS. Notes.*

[3] Seebohm's *British Birds*, vol. ii. p. 178.

THE SAND MARTIN.

SAND SWALLOW, BANK SWALLOW.

Cotile riparia.

The Bitterbank.

The Swallow tribes in spring
Thus daily multiply upon the wing,
As if the air, their element of flight,
Brought forth new broods from darkness every night.
MONTGOMERY.

THE arrival of the Sand Martin, which is smaller and more sombre in colour than its congeners, may be looked for about the same time as that of the Chimney Swallow, but it sometimes comes a little earlier.[1] If the weather be cold, it hawks about the sheltered reaches of the Whitadder and other streams for some days after its return, the neighbourhood of Allanton Bridge being a favourite resort, and there it is often seen for the first time in the season.

It seldom frequents buildings,[2] but nests together in large colonies in the banks of streams, the vertical faces of sand-pits and quarries, and steep railway embankments, choosing situations where the soil is of a sandy, friable nature, and in which it can easily bore more or less

---

[1] Dr. Stuart records seeing a pair at Allanton Bridge on 24th March 1884.—*Hist. Ber. Nat. Club*, vol. x. p. 576.

[2] A pair nested in a hole in the wall round the policy-ground at Paxton in the summer of 1881.

horizontal holes of sufficient depth for the nest. These are of various lengths, extending from 18 inches to 3 feet; and at the end, which is somewhat enlarged, the nest, consisting of dry grass and a few feathers, is built. The eggs, which are from four to six in number, are pure white. Mr. Hardy mentions the following amongst the numerous breeding-places of the Sand Martin in the county :—Near Redheugh; on the coast near Billsdean; in the cliff near Saltpan-hall; on the Whitadder at East Blanerne; on the Eye below Quixwood; near Butterdean; sand-pit at Horseley road; on Heriot Water below Stockbridge; above Tower Wood; and quarry at Paddock Cleuch. It likewise nests in a sand-pit at Paxton; on the Tweed at Sunnyside, near Milne Graden; on the sandy banks of the Dye, between the junction of that stream with the Whitadder and Dye Cottage; on the banks of the Blackadder, at Black Castle Rings; and at Caldra.

The food consists entirely of insects, to secure which the bird frequents the neighbourhood of streams.

It leaves earlier in the autumn than the Chimney Swallow and the Martin, and very little is known regarding its winter quarters.

THE TREE-CREEPER.

CREEPER, TREE-SPEELER, BROWN WOODPECKER.

Certhia familiaris.

𝕿𝖍𝖊 𝕭𝖆𝖗𝖐-𝕾𝖕𝖊𝖊𝖑𝖊𝖗.

The greenwood for him is the place of his rest,
And the broad-branching tree is the home he loves best.
. . . .
There up the tree-trunk, like a fly on the wall,
To pick the grey moss, runs the Tree-Creeper small.
MARY HOWITT.

ALTHOUGH the Tree-Creeper, which is one of our smallest birds, is to be found in all our woods and plantations where the trees are large, at every season of the year, it is not much noticed; for its sombre plumage and retiring habits do not attract attention, and when seen creeping up the trunk of a tree at a little distance it is more like a mouse than a bird. It generally alights at the bottom of the trunk near the ground, and creeps upwards by short jerks, in a somewhat winding manner, diligently searching every crevice of the bark of the trunk and larger branches for insects, and frequently uttering a low chirp. Having finished its work on one tree, it quickly flies down, curving as it nears the ground, to the root of another, and immediately begins its spiral ascent. Its claws are curved and sharp, and the tail feathers are stiff and pointed. In climbing, the bird crouches close to the tree, keeping its tail pressed against the bark. Its song, which is but seldom heard, is said to resemble that of the Golden-Crested Wren.

Its nest is generally found in a crevice of the bark of a decayed tree, though it also chooses other situations.[1] The nest is usually formed of dry grass and moss, lined with feathers; and the eggs, which are from six to nine in number, are white, marked with brownish red spots.

---

[1] Mr. A. H. Evans, Cambridge, discovered a nest some years ago, under the tiles on the roof of the boat-house, in the policy-ground of Paxton House.

THE GOLDFINCH.

GOLDSPINK, GOUDSPINK, GOLDIE.

Carduelis elegans.

𝕿𝖍𝖊 𝕲𝖔𝖑𝖉𝖎𝖊, 𝕿𝖍𝖊 𝕲𝖔𝖔𝖑𝖉𝖎𝖊, 𝕿𝖍𝖊 𝕲𝖔𝖔𝖑𝖉𝖘𝖕𝖎𝖓𝖐.[1]

About an bank with balmy bewis

With gallant Goldspinks gay,
The Mavis, Merle, and Progne proud,
The Lintquhyt, Lark, and Lavrock loud,
Salutit mirthful May.

ALEX. MONTGOMERY, *The Cherrie an' the Slae,*
1567-1603.

IN former times, when thistles, burdocks, and other weeds were much more abundant in the county than they are at present, the Goldfinch was frequently seen in small flocks feeding on their seeds, but now it is a comparatively rare bird.

Mr. Hardy mentions that about 1793, when his father was a boy, great numbers frequented the top of the seabanks at Oldcambus, where they fed on the seeds of the burdock that used to be very abundant there.[2] He has also informed me that they were said to have been numerous long ago about Butterdean Mill and Oldcambus West Mains,

---

[1] Boys in some districts of the county repeat the following rhyme :—

A Blackie and a Blackbird,
A Laverock and a Lark,
A Goldie and a Goldspink,
How many birds is that?

[2] *Hist. Ber. Nat. Club*, vol. vii. pp. 296, 297.

and that he saw a small flock at Penmanshiel during a snow-storm towards the end of December 1834. Mr. John Wilson, late of Edington Mains, has told me that between 1820 and 1830 he used to see considerable numbers about the tall trees on that farm. Mr. W. Duns, builder, Duns, states that when the castle was being built there in 1821, a Goldfinch nested in a big plane-tree near its site, and that they were so plentiful at that time in the neighbourhood that some men about Duns made their living by catching them with bird-lime between harvest and spring, and selling them as cage-birds. He remembers that an English bird-catcher who then lived in Duns, and who was known by the "nickname" of "Bird Jock," caught great numbers of them, and had sometimes as many as thirty or forty in his house at one time. Mr. John Aitchison, plasterer, Duns, says that long ago Crunkley was a great resort of "Goldies," for "mother of earth," or chickweed (*Stellaria media*) grew there plentifully; also that Buxley Dean, between Manderston and Manderston Mill, was much frequented by these birds, for it abounded with thistles, horse-knots (*Centaurea nigra*), and other weeds. Mr. Peter Scott, Lauder, relates that Goldfinches were common in Lauderdale about forty-five years ago, and that their nests were sometimes found about that time in the grounds of Thirlestane Castle. The rocky deans near the sea-coast at Lamberton are also said to have been frequented by "Goldies" in former times, and they were likewise found about Fairneyside. Mr. Hardy records that a flock of "this now rarely seen bird" visited the neighbourhood of Cockburn Law about the end of November 1874, and that, several years previous to that time, he saw a flock near Oldcambus picking thistle seeds.[1] According to Mr. Kelly a flock of eighteen Gold-

---

[1] *Hist. Ber. Nat. Club*, vol. vii. p. 296.

finches was seen at East Mains, Lauder, in December 1874, and several flocks during the winter of 1874-75.[1] In the same winter they visited the sunny braes of Old Thirlestane,[2] and three were seen in the previous October on the hill above Preston.[3] Several were observed at Thornydykes near Westruther, by Mr. John Ferguson, Duns, on the 9th of February 1877;[4] and the Rev. W. Stobbs records the occurrence of a pair about Gordon in the summer of 1880.[5] Mr. Lockie, Spottiswoode, informs me that they are occasionally seen in the vicinity of Legerwood Loch. Mr. John Thomson, Maxton, writes to me that in January and February 1886, several flocks were seen at Milfield and the Willow Bogs near Mertoun. The banks of the Whitadder, near Edrington Mill, are still occasionally frequented by these beautiful birds; several having been caught there in December 1886, by Mr. John Edgar, Paxton, with a call-bird and lime twigs stuck round its cage. Miss Georgina Milne-Home, Milne Graden, procured two pairs of Goldfinches from the south of England in the spring of 1886, and, for the purpose of trying if they would remain and nest, she set them at liberty in the policy-grounds there, shortly after their arrival. This laudable attempt to introduce the bird to that neighbourhood is said to have been successful.

The nest is very neat, and somewhat resembles that of the Chaffinch, being, like that of the latter bird, often built

> Fast to the blushing apple's forked branch,
> Amid the blossom of the codlin tree.
>
> HURDIS.

---

[1] *Hist. Ber. Nat. Club*, vol. vii. p. 289.
[2] *Ibid.* vol. vii. p. 521.
[3] *Ibid.* vol. viii. p. 494.
[4] *Ibid.* vol. viii. p. 192.
[5] *Ibid.* vol. ix. p. 230.

THE GOLDFINCH.

> Sometimes suspended at the limber end
> Of plane-tree spray, among the broad-leaved shoots,
> The tiny hammock swings to every gale.
> GRAHAME.

The eggs, which are four or five in number, are white, with a few spots and streaks of purple- and dark reddish brown.

THE SISKIN.

ABERDEVINE.

Carduelis spinus.

The Siskin.

*Beautiful birds! ye come thickly around,
When the bud's on the branch and the snow's on the ground.*

COOK.

THIS pretty little bird generally makes its appearance in the county about the month of October, and leaves us for the north again [1] in March and April.

It chiefly frequents localities where alder and birch trees abound, such as the margins of our streams, ponds, and bogs; and small flocks may be sometimes seen, accompanied by Redpolls, feeding on the seeds of these trees, which form their favourite food. While thus occupied, it assumes a variety of interesting attitudes like the Titmouse; and when a flock is feeding, the little twittering birds are frequently so eagerly engaged in searching for food as to allow a person to walk close up to the tree or bush upon which they are feasting, without showing any fear of his approach. Of this fearlessness the bird-fanciers in some districts take advantage, and use a long, slender wand, like a fishing-rod,

---

[1] The Siskin has been observed on migration in *Autumn* 1881.—Isle of May, 4th and 5th Oct. *Spring* 1882.—Isle of May, with other Finches, from 3rd March to 22nd May. *Autumn* 1882.—Isle of May, 7th Aug. to 16th Oct. *Autumn* 1885.—Isle of May, 30th Oct. [I see no record of the Siskin at the Farnes.]—*Reports on the Migration of Birds*, 1879-85.

with its point tipped with bird-lime, to catch Siskins by touching them with it, when their feathers adhere to the lime with which it is smeared.

This species does not generally visit the same parts of the county every year, but frequently appears in considerable numbers in a locality where it is not seen again for several successive seasons. As an instance of this peculiarity, it may be mentioned that, in the winters of 1870-71 and 1874-75, several flocks frequented the birch-trees in the policy-grounds of Paxton House, where none have been observed for the last ten years. It seems to occur in the neighbourhood of Duns Castle frequently, and as it is reported to have been seen in the woods there during summer, it is possible that it may have bred in that locality. There is, however, no record of the nest having been found in Berwickshire.

The following places in the county, in addition to those above mentioned, are recorded as haunts of the Siskin:— Alders at Clockmill, near Duns;[1] wood at the side of the Luggy in Lauderdale;[2] sides of Tower Burn, near Pease Mill;[3] banks of the Blackadder;[4] Gordon Moss and Fans;[5] and Edington Mill.[6] I have likewise notes of its appearance near Mertoun; at the Langton Burn, near Langton House; and on the banks of the Leader, near Earlston.

---

[1] *Hist. Ber. Nat. Club*, vol. vii. p. 121.
[2] *Ibid*. vol. vii. p. 303.
[3] *Ibid*. vol. vii. p. 513.
[4] *Ibid*. vol. vii. p. 521.
[5] *Ibid*. vol. ix. pp. 230, 561.
[6] *Ibid*. vol. x. p. 573.

THE GREENFINCH.

GREEN LINNET, GREEN GROSBEAK, GREEN LINTIE.

Coccothraustes chloris.

The Green Lintie.

*I love the broom where the gay Green Linnet
Bigs its wee bower on the broken tree.*
H. SCOTT-RIDDELL, *The Bonny Broom.*

The Grene serene sang sueit, quhen the Goldspynk chantit.
SIR DAVID LINDSAY, *Complaynt of Scotland.*

THE Greenfinch is a very common bird in the county, and may be seen in numbers wherever there are trees and hedges. It remains with us all the year, but our residents may probably receive additions to their numbers in autumn from migratory flocks from the north.[1]

Towards that season it assembles in flocks with Chaffinches and other birds, and frequents the stubble fields, where it feeds on seeds of various kinds, as well as on the waste corn. When severe weather occurs, with frost and snow, it is seen in great flocks in stackyards, with House Sparrows, Yellow Buntings, and Chaffinches, feeding round the stacks, about the barn doors, or on grain thrown down

---

[1] The Greenfinch has been regularly observed on migration at the Isle of May in autumn and spring. *Spring* 1881.—4th March, 14th April. *Spring* 1882.—17th and 18th March. *Autumn* 1882.—9th to 23rd Oct. *Spring* 1883.—17th Feb. to 5th April. *Autumn* 1883.—1st to 3rd Nov. *Spring* 1884.—10th Feb. to 25th March. *Autumn* 1884.—4th Nov. to 26th Jan. 1885. *Spring* 1885.—Feb., March, and April. *Autumn* 1885.—5th Nov.—*Reports on the Migration of Birds,* 1879-85.

for the poultry. Shortly after a stack has been thrashed, in snowy weather, the place where it stood is often seen quite covered with Greenfinches, all eagerly searching for food. A favourite resort for the great flocks of these birds which frequent our farm-steadings in winter is the sloping thatch on the top of some of the ricks, and this has been pleasingly referred to by Grahame in the following lines :—

> Pleasant the path
> By sunny garden wall, when all the fields
> Are chill and comfortless; or barnyard snug,
> Where flocking birds, of various plume, and chirp
> Discordant, cluster on the leaning stack.

As spring advances, the Greenfinch betakes itself to our plantations and hedges, where it builds its nest, composed of fibrous roots, moss, and wool, lined with horse-hair and feathers. The eggs, which are generally five or six in number, are white, spotted and blotched, chiefly at the larger end, with dark reddish brown.

The food consists of grain, seeds,[1] and insects.

---

[1] Mr. Hardy, writing on the effects of the winter of 1878-79, mentions that during the severe weather then experienced Greenfinches attacked the heads of burdocks growing in neglected places, and dismembered them for food.—*Hist. Ber. Nat. Club*, vol. ix. p. 124.

THE HOUSE SPARROW.

COMMON SPARROW, SPROUG.

Passer domesticus.

𝕿𝖍𝖊 𝖘𝖕𝖆𝖗𝖗𝖞.

From the summit of the leafless elm
Excessive chirpings pour ; fond parliament,
Where all are speakers, and none sits to hear.
The Sparrow-couple with industrious bill
The scatter'd straws collect, contriving snug,
Under the cottage-eave or low-roof'd barn,
Their genial couch.

HURDIS.

ALTHOUGH this familiar bird is numerous about all the farmsteadings and villages in the county, it has, fortunately, not yet multiplied to such an alarming extent with us, as to render its depredations in the corn field, like those of the wood pigeon in former years, a subject of discussion at agricultural meetings. It has, however, of late years, increased so much in various parts of England, and done so much damage to the crops, that the farmers in various districts there have formed special clubs for the destruction of the Sparrow, and offered, in some instances, a reward of sixpence for each dozen of heads produced.[1]

Mr. J. H. Gurney, jun., Northrepps, Norwich, a well-known ornithologist, has lately written an exhaustive account of the House Sparrow, from a farmer's point of view,[2] in which he says that "No one can for a moment

---

[1] *The House Sparrow*, by J. H. Gurney, jun.; W. Wesley and Son, London, 1885.   [2] *Ibid.*

doubt that the Sparrow question is now a very important one, and that it is becoming year by year more so ; that is, if a tithe of what is said and written about it at farmers' clubs and in the agricultural newspapers be true." He has been so good as to give a table showing the results, during each month of the year, of 694 dissections of House Sparrows' crops, made by various hands, in various places ; adding that to give a summary of the table in a few words it may be said that about 75 per cent. of an adult Sparrow's food during its life, is corn of some kind, and that the remaining 25 per cent. may be roughly divided as follows :—Seeds of weeds, 10 per cent. ; green peas, 4 per cent. ; beetles, 3 per cent. ; caterpillars, 2 per cent. ; insects which fly, 1 per cent. ; other things, 5 per cent. He likewise states that in young Sparrows, up to the age of sixteen days, not more than 40 per cent. is corn, while about 40 per cent. consists of caterpillars, and 10 per cent. of small beetles.[1] Notwithstanding the various opinions which have been held as to the usefulness, or otherwise, of the House Sparrow to the farmer, we must now accept the results of the 694 dissections above mentioned, as conclusive evidence on the subject, for, as our national poet says—

> Facts are chiels which winna ding,
> And downa be disputed.

The House Sparrow often works a considerable amount of mischief in our gardens by destroying the blooms of two of our most beautiful spring flowers—crocuses and primroses ; and it also eats young peas when sprouting through the ground, and green peas from the pod. Every year in spring the primroses in my garden are completely destroyed by these birds nipping through the blooms when in full beauty, at the part where the seed-pod forms, and thus it

---

[1] *The House Sparrow*, by J. H. Gurney, jun., p. 17.

often happens that where thousands of lovely primroses may be seen in full bloom in the evening, the ground around the plants on the following morning is found strewed with their withering petals, the Sparrows having been at their mischievous work soon after sunrise. They simply nip through the incipient seed-pod at the base of the petals, and let the flower fall to the earth.

It is likewise a great enemy to two of our favourite birds—the Martin and the Swallow—for it takes possession of their nests in summer, and drives the birds away from their haunts. As an instance of this it may be mentioned that, for several years past, three or four pairs of Swallows have nested under the eaves of my house, and it was a source of great pleasure in the summer evenings to see them flying round the building and constantly visiting their nests, as well as to hear their delightful twitterings early in the morning. In 1886, however, they were so much annoyed by the House Sparrows continually taking possession of their nests and fighting with them, that, unfortunately, they all left with the exception of a single pair.

Although its general character is undoubtedly bad, yet this species has some redeeming points, for it is a cheerful, familiar bird. Its chirping is heard about the windows of our houses early in the cold winter mornings, when deep snow covers the ground, and all the small birds are silent; and its "parliaments," which are often held in hedges near stackyards, remind us of fine sunny days when

> Fair-handed Spring unbosoms every grace ;
> Throws out the snowdrop and the crocus first ;
> The daisy, primrose, violet darkly blue,
> And polyanthus of unnumber'd dyes.
>
> THOMSON, *Spring*.

It has been occasionally observed on migration at the

lighthouses on the Isle of May and the Farne Islands, in spring and autumn.[1] The well-known nest of the Sparrow is large, domed, and generally composed of straw, hay, and dry grass, lined with a profusion of feathers. Mr. Hardy has been so good as to send me a note about two nests at Penmanshiel, out of which he counted no fewer than 742 feathers, besides sundry patches of hair from the backs of cattle, thistle-down, and tufts of wool.[2] The nest is built in a great variety of situations, a favourite spot being amongst the branches of roses, or other climbing plants trained up the walls of houses.[3] The eggs are usually five or six in number, of a greenish white, blotched and spotted with ash colour and dark brown.

Pure white and piebald specimens of this bird are sometimes seen in the county.

---

[1] It was observed in *Spring* 1882 at Isle of May, 3rd March to 22nd May. *Autumn* 1882.—Isle of May, October; Farnes, 2nd Nov. *Note* by lighthouse-keeper: "Seldom seen here." *Autumn* 1884.—Isle of May, November to December. *Spring* 1885.—Isle of May.—*Reports on Migration of Birds*, 1879-85.

[2] In another note about the Sparrow, Mr. Hardy says: "*5th August* 1858.—Barley is said to be already as hard in its progress towards ripening, that one *could shoot Sparrows with it.*"

[3] A very large colony of Sparrows breed in the walls of the old tower of Evelaw, near Wedderlie, harbouring in the ivy which covers the eastern gable.

THE TREE SPARROW.

MOUNTAIN SPARROW.

Passer montanus.

The Sparrow chirmis[1] *in the wallis clyft.*
                              GAVIN DOUGLAS, *Virgil.*

THE Tree Sparrow is found in small numbers round the coast of Berwickshire, at various places between Lamberton and St. Abb's Head, and has likewise been observed in several localities in the interior of the county. Mr. George Bolam records that he saw several pairs, and found three nests with young, at St. Abb's Head, in June 1878, and that about the same time he noticed some birds of this kind at Reston Station.[2] Mr. W. Evans, Edinburgh, has informed me that, on the 22nd of April 1886, he saw a Tree Sparrow on the old tower by the roadside, a little distance to the west of the Pease Bridge, and that, on the 8th of June following, he noticed a pair dusting themselves on the bank of the small ravine immediately under the walls of Hutton Hall, on the Whitadder.

This species has been frequently seen on migration at the Isle of May in spring and autumn.[3]

It places its nest, which is more carefully built and contains a less quantity of materials than that of the House Sparrow, in various situations, such as in crevices of rocks; under the tiles on the roofs of buildings, in holes in trees

---

[1] Chirps.   [2] *Hist. Ber. Nat. Club*, vol. viii. p. 495.
[3] *Reports on the Migration of Birds*, 1879-85.

THE TREE SPARROW.

or banks, and sometimes under the coping of old walls. The eggs, which are from four to six in number, are very like those of the House Sparrow, but not so large. The Tree Sparrow is a smaller bird than the House Sparrow, from which it may be distinguished by its chestnut head, its black ear coverts surrounded by white, and by the double bar of white on its wing.

THE CHAFFINCH.

SHILFA, SHEELY, SHELLY, SHELL-APPLE, SPINK, SCOBDY, SKELLY, PINK, TWINK, BEECH-FINCH, HOOSE-FINCH.

Fringilla cœlebs.

The Shilfa, The Shulfie, The Upper Shealer.[1]

> At such a still and sultry hour as this,
> When not a strain is heard through all the woods,
> I've seen the Shilfa light from off his perch,
> And hop into a shallow of the stream.
> Then, half afraid, flit to the shore, then in
> Again alight, and dip his rosy breast
> And fluttering wings, while dew-like globules coursed
> The plumage of his brown empurpled back.
>
> GRAHAME, *Birds of Scotland.*

THE Chaffinch is one of our commonest birds, and may be seen in all our shrubberies, woods, and hedgerows, during spring and summer. Towards autumn it congregates in flocks, and haunts the stubble fields with Greenfinches and other birds, where it picks up fallen grain and seeds. Males are generally as numerous as females in the assemblages seen here at that time, although in some parts of the country the sexes have been observed to keep separate in autumn and winter. When feeding together in numbers Chaffinches generally fly down more or less gradually from the adjacent buildings or trees; and then, on any alarm, they mount up in

---

[1] Mr. Hardy says this name is used in the Merse, being apparently a corruption of the Northumbrian name "Apple-Sheeler."

a body to some elevated position, again proceeding to their feeding ground, a few at a time as before, after the alarm is past. The Chaffinch continues to haunt the stubble fields during open weather in winter, but when snow occurs it frequents farm-steadings and stackyards with Sparrows and Linnets, which help themselves to the food thrown down to poultry, or pick up grain round the sides of the stacks and about barn-doors and cattle-courts. When a Sparrow Hawk approaches the steading at this season, this bird's loud alarm-note—"pink, pink, pink"—is generally the first indication of his presence; and the same note is amongst the loudest when small birds are engaged in persecuting an Owl with their cries, on discovering him in his retreat in the woods during the day-time. If very severe weather sets in, with much frost and snow, the Chaffinch appears to leave the more exposed parts of the county for sheltered localities near the sea-coast. In summer it subsists almost entirely on insects, many of which are injurious to the farm and garden. It roosts at night in evergreens and young fir woods during autumn and winter.

Mr. Hardy says that in Berwickshire when the "weet, weet, weet" of this bird is much heard, it is considered to be an indication of snow or rain. The Chaffinch has been frequently observed on migration [1] at the Isle of May

---

[1] *Autumn* 1879.—Farnes, 28th Sep. and 1st Oct. *Autumn* 1880.—Isle of May, 19th Oct.; Farnes, 1st Sep. *Spring* 1881.—Isle of May. *Spring* 1882.—Isle of May, 17th and 18th March. *Spring* 1883.—Isle of May, 2nd March and in April. *Autumn* 1883.—Farnes, 29th and 30th Nov. *Spring* 1884.—Isle of May, 12th Feb. and 3rd to 10th May. *Autumn* 1884.—Farnes; great numbers 16th Oct. and 2nd Nov. Mr. Cordeaux, the Reporter for the East Coast of England, remarks with reference to the migration of the Chaffinch at this time:— "The migration is extensive, and covers the whole of the east coast. Through September and October immense numbers are recorded as crossing—the first flocks consisting of young of both sexes and old females; old cocks later." Isle of May, 25th to 27th Sep. *Spring* 1885.—Isle of May. single records in Feb., March, April, and May. *Autumn* 1885.—Isle of May, 14th Oct. and 14th Nov.—*Reports on the Migration of Birds*, 1879-85.

and the Farne Islands, and probably our Berwickshire birds may receive additions in autumn from migratory flocks.

The male in his nuptial dress is a handsome and beautiful bird, and the colour of the cheek of beauty has been compared to that of his breast—

> Her cheek is like the Shilfa's breast,
> Her neck is like the Swan's.
> MARY STEWART, *Hist. Drama*, p. 113.

The joyous song of the Chaffinch is heard when the earliest primroses appear in our gardens, and when the apple is in blossom.

> . . . Full and clear the sprightly ditty rings,
> Cheering the brooding dam: she sits concealed
> Within the nest deep-hollowed, well disguised
> With lichens grey and mosses gradual blent,
> As if it were a knurle in the bough.
> GRAHAME.

The eggs are from four to six in number, and are generally of a light bluish green, clouded with pale reddish brown, and spotted with dark reddish brown.

Several examples with white or abnormal plumage have occurred in Berwickshire of late years. Mr. Robert Renton mentions that a light cream-coloured bird was shot at Fans on the 1st of January 1883, and that in 1881 two of a somewhat similar colour were observed there.[1] A white Chaffinch was seen near Swinton in 1883, and Dr. Stuart records that a few years before that time one of the same colour was noticed in a strip of wood near Allanton Free Church.[2] Mr. Peter Cowe, Lochton, lately showed me a specimen killed there in 1871, with head and under-parts

---

[1] *Hist. Ber. Nat. Club*, vol. x. p. 572.   [2] *Ibid.* vol. x. p. 573.

white, wings dark and mottled with white, tail dark, and breast tinged with cream colour.

Mr. Hardy says that the Chaffinch appears to be a long-lived bird, for a Mr. Moffat, a retired schoolmaster in Ayton, had one in 1845 which had lived as a cage-bird for eighteen years.

THE BRAMBLING.

MOUNTAIN FINCH, BRAMBLE FINCH.

Fringilla montifringilla.

The Cock o' the North.

When biting Boreas, fell and doure,
Sharp shivers through the leafless bow'r;
When Phœbus gies a short-lived glow'r
Far south the lift,
Dim-dark'ning through the flaky show'r,
Or whirling drift.

BURNS.

THE "Cock o' the North" is mostly associated in our minds with the severe weather of winter, when deep snow covers the level expanse of the Merse, and curlers are on the ice; for then it is seen in small numbers about the stackyards, feeding with the Chaffinches and Linnets on any waste grain or seeds which it can find.

It is an autumn and winter visitor to most parts of the county, coming from the north in October and November, and returning northwards again in March and April.[1] Although the Brambling is seen in Berwickshire every year, it is much more plentiful in certain districts in some years than in others, its visits being somewhat erratic and uncertain, so far as particular localities are concerned.

When the weather is mild it frequents the open fields with flocks of Chaffinches and other birds, and as it

---

[1] Mr. George Bolam records seeing one in his garden at Berwick-on-Tweed in the spring of 1884 as late as the 1st of May.—*Hist. Ber. Nat. Club*, vol. x. p. 588.

resembles the male Chaffinch in manners, appearance, and plumage, it can scarcely be distinguished at a little distance when feeding amongst them. It likewise resorts to plantations, where it feeds upon the beech-mast. Large flocks were seen in the beech woods at Paxton in the autumn of 1874, but they nearly all disappeared when the heavy snow-storm of December of that year came on, only a few remaining about the farm-steadings in the neighbourhood during the severe weather which followed. In 1875 none visited Paxton, in 1876 they were very numerous, in 1877 only a few were noticed, and none in 1878. Mr. Kelly mentions that, at the farm of Pilmore in Lauderdale, Bramblings have been observed feeding among the lint refuse left after thrashing.[1] He also records its appearance in winter at Lylestone, Newmills, and Threeburnford near Oxton.[2] Mr. Renton states that it was seen at Fans, on the 10th of January 1881, during hard frost and snow ; and Dr. Stuart mentions it as occurring at East Blanerne on the 4th of February of the same year.[3] Mr. Lockie writes that it is often noticed about Legerwood, Westruther, and Spottiswoode, and also in the Earlston district, during the winter months. Mr. Hardy notes that it comes to the stackyards at Penmanshiel and Oldcambus in severe weather, but not in large numbers ; and that it is usually ready to resort to a horse-hair " hoop girn " during snow.

There are several records of the Brambling being seen on migration at the Farne Islands and the Isle of May in autumn and spring, the earliest appearance in autumn which is recorded being at the Farne Islands on the 21st of September 1882, and the latest in spring, at the Isle of May, on the 7th of April 1885.

---

[1] *Hist. Ber. Nat. Club*, vol. vii. p. 303.　　[2] *Ibid.* vol. viii. p. 144.
[3] *Ibid.* vol. ix. p. 555.
[4] *Reports on the Migration of Birds*, 1879-85.

THE LINNET.

GREY LINNET, BROWN LINNET, ROSE LINNET, WHIN LINNET, GREATER REDPOLL, LINTIE, LINTWHITE.

Linota cannabina.

𝕿𝖍𝖊 𝕲𝖗𝖊𝖞 𝕷𝖎𝖓𝖙𝖎𝖊, 𝕿𝖍𝖊 𝕽𝖔𝖘𝖊 𝕷𝖎𝖓𝖙𝖎𝖊, 𝕿𝖍𝖊 𝖂𝖍𝖎𝖓 𝕷𝖎𝖓𝖙𝖎𝖊.

> *The Warblers are heard in the grove,*
> *The Linnet, the Lark, and the Thrush,*
> *The Blackbird and sweet cooing Dove,*
> *With musick enchant ev'ry bush ;*
> *Come, let us go forth to the mead,*
> *Let us see how the primroses spring ;*
> *We'll lodge in some village on Tweed,*
> *And love while the feather'd folks sing.*
>
> CRAWFORD, *Tweedside.*[1]

> *The lyntquhit sang cunterpoint quhen the oszil zelpit.*
>
> SIR DAVID LINDSAY, *Complaynt of Scotland.*

IN former times, before the advance of agricultural improvements had taken place to such a great extent in the county as now, and when on many farms in the Merse there still remained some waste uncultivated ground covered with whins and broom, the Linnet was much more plentiful than it is at present. It is, however, even yet, numerous in suitable localities, such as the neighbourhood of wild grounds and moorlands, rocky deans, sea-banks, and the like; where whins, with burdocks and horse-knots, abound. The banks

---

[1] This fine Scottish song was written in 1731 by Robert Crawford of Auchinames.

of the Leader seem at one time to have been a favourite haunt of this bird, for it is introduced in the beautiful Scottish song of "Leader Haughs and Yarrow"[1]—

> A mile below, wha lists to ride,
>   They'll hear the Mavis singing,
> Into St. Leonard's banks she'll bide,
>   Sweet birks her head o'er-hinging :
> The Lintwhite loud and Progne proud,
>   With tuneful throats and narrow,
> Into St. Leonard's banks they sing
>   As sweetly as in Yarrow.

It is a great favourite with bird-fanciers in the county as a cage-bird, on account of its pleasing song; and has been known to live with one of the fraternity, in Ayton, for ten years. It is usually taken with bird-lime or girns made of horse's hair, but it is often reared from the nest. Boys have been known to "pudge" the young ones by placing a kind of framework of small sticks round and over the nest to keep them from flying out, the old birds continuing to attend to them until they can feed themselves, when they are removed from the nest and put into a cage.

Mr. Hardy writes to me that, owing to the lively disposition of this bird, it is usual for the country people of Berwickshire, when alluding to a cheerful person, to remark that he, or she, is as "canty as a lintie"; likewise that when many young Linnets congregate and sing together towards the end of the year, it is augured that bad weather is at hand.

Towards autumn the Linnet assembles in flocks and frequents the stubble, pasture, and fallow fields, in search of seeds of various kinds, especially those of the wild mustard

---

[1] This song, written by Nicol Burne, is given with the music in Ritson's *Scottish Songs*, second edition, vol. ii. p. 458.

(*Sinapis arvensis*),—called "Shirts" in Berwickshire,—and of the Runch (*Raphanus raphanistrum*). In waste and uncultivated places it finds the seeds of the burdock (*Arctium*) and knap-weed or horse-knops (*Centaurea nigra*); and in summer those of the dandelion form part of its food. Great numbers are attracted to fields of ripening turnip-seeds.

In ordinary seasons the Linnet remains with us throughout the year, but when very severe weather sets in, such as we experienced in the terrible winters of 1878-79 and 1880-81, it usually leaves the county for a milder climate. It has been observed on migration at the Isle of May in spring and autumn,[1] when probably our Berwickshire Linnets may receive additions to their numbers from migratory flocks. It was observed to be scarcer in the county for some time after the severe winters above mentioned.

On the arrival of settled weather in spring the birds which have kept together in flocks during the winter, break up into scattered pairs, and—

> When whiny braes are garlanded with gold,
> And blithe the lamb pursues in merry chase
> His twin around the bush; the Linnet then,
> Within the prickly fortress, builds her bower,
> And warmly lines it round with hair and wool
> Inwove.
>
> GRAHAME, *Birds of Scotland.*

On the 21st of June 1888, Mr. Compton Lundie of Spital and I found several nests of this species in isolated whin bushes on the rough pasture ground at St. Abb's Head. They contained eggs, which varied from four to five in

---

[1] The Linnet has been observed on migration at the Isle of May:—*Spring* 1880.—18th May. *Autumn* 1881.—18th Aug. *Spring* 1882.—3rd March and 22nd May. *Autumn* 1882.—23rd Aug. to 8th Dec.; vast rush, 9th to 23rd Oct. *Spring* 1884.—25th March. *Autumn* 1884.—25th to 27th Sep.—*Reports on Migration of Birds,* 1879-85.

number and were white, tinged with green or blue, spotted, speckled, or blotched with reddish brown.

When in full summer plumage a fine old male has the fore-part and top of the head glossy blood-red. In this dress the bird is called a "Rose Lintie" in Berwickshire.

THE LESSER REDPOLL.

REDPOLL, SMALLER REDPOLL, LINNET.

Linota rufescens.

𝕿𝖍𝖊 𝕽𝖊𝖉𝖕𝖔𝖑𝖑.

See the birds together,
In this splendid weather,

And each feathered neighbour
Enters on his labour.
Sparrow, Robin, Redpoll, Finch, Linnet, and the Wren.

MARY HOWITT, *An April Day.*

IN Berwickshire the Lesser Redpoll is generally seen in small flocks with Siskins, feeding upon the seeds of birch and alder trees during the autumn and winter months, but it is also observed in small family parties towards the end of summer. Of late years some small flocks of five or six have several times been noticed by me in the latter part of July frequenting the young fir plantations near my house at Paxton, being apparently on local migration. A pair were seen in the beginning of June 1887 haunting the same young woods, where they appeared to be nesting; but on a very careful search being made, no nest could be found, and the birds shortly afterwards left for other quarters. The great majority of the Redpolls which form the flocks seen in the county in winter, appear to come in autumn, and leave in spring.[1]

---

[1] The Lesser Redpoll was observed on Migration:—*Autumn* 1881.—Isle of May, 20th and 24th Oct.; Farnes, 26th Oct. *Spring* 1882.— Isle of May. *Autumn* 1882.—Isle of May. *Autumn* 1885.—Isle of May.—14th Oct.; also 23rd Jan. 1886.—*Reports on the Migration of Birds,* 1879-85.

Mr. Lockie writes to me that this species breeds every year in the neighbourhood of Corsbie Tower, Spottiswoode, and Hydsidehill. It constructs a neat nest, composed of twigs, dry grass, moss, and wool, lined with vegetable down, such as from the catkins of the willow, and sometimes with feathers and hair. The eggs, which are from four to six in number, are generally greenish blue marked with reddish brown.

The Lesser Redpoll is recorded in the *History of the Berwickshire Naturalists' Club* as visiting the following places in the County :—Near Allanton ;[1] wood at Luggy ; Lauderdale ;[2] Penmanshiel ;[3] Duns Castle woods ;[4] Old Thirlestane ;[5] Gordon Moss ;[6] and Edington Mill.[7] I have likewise notes of its occurrence near the Grange Wood, in Coldingham parish ; and on the Leader below Rhymer's Mill, near Earlston. Several Redpolls were caught while feeding on the alders by the side of the millpond at Nabdean near Paxton, in December 1887, the bird-catchers using a long slender stick, smeared at the point with bird-lime to take them, in the same way as they capture Siskins.

There is no record of the occurrence of the Mealy Redpoll (*Linota linaria*) in Berwickshire. Mr. Seebohm considers this and the Lesser Redpoll as forms of the same bird.[8]

---

[1] *Hist. Ber. Nat. Club*, vol. vii. p. 284.
[2] *Ibid.* vol. vii. p. 303.
[3] *Ibid.* vol. vii. p. 513.
[4] *Ibid.* vol. vii. p. 513.
[5] *Ibid.* vol. vii. p. 521.
[6] *Ibid.* vol. ix. p. 561.
[7] *Ibid.* vol. x. p. 573.
[8] Seebohm, *Brit. Birds*, ii. 115.

THE TWITE.

MOUNTAIN LINNET, HEATHER LINTIE.

Linota flavirostris.

The Heather Lintie.

Sweet amang the knowes and brakens
Sing the Linties, chanting clear;
Sweet amang the bogs and mosses,
Flora's gems the loneness cheer.
     DR. HENDERSON.

THE Heather Lintie, which is very like the Common Linnet, but not so large, is occasionally seen in small flocks in the lower parts of the county during winter, and in summer it is found in the Lammermuirs. In the end of December 1874, a small flock visited the Crow-Dean Wood at Paxton, where they fed on the seeds of the avens (*Geum urbanum*), which is plentiful there. Mr. Hardy mentions having seen this bird some years ago early in October, in the woods round the "Dog-bush" near Marygold,[1] and also that old shepherds in the Lammermuirs used to know its nest. Mr. W. Duns, builder, Duns, has informed me that about sixty-five years ago, when he was a boy, his father sometimes brought home Heather Linties in a "stocking leg," to be kept as cage-birds. They were called "Peepers," and were caught with horse-hair "girns" at Borthwick, on the march between Duns Castle and Langton estates. Mr. Mason,

---

[1] *Hist. Ber. Nat. Club*, vol. v. p. 233.

Auchencraw, has told me that he used to take numbers of these birds some years ago, in winter, at West Greenfield. It is said to frequent Coldingham Moor, and the Rev. George Cook, Longformacus, writes to me that Mr. Smith has frequently seen it at Whitchester, and has also found its nest there. On the 10th of March 1887, I saw a small flock feeding under the birch-trees which adorn the banks of the Dye near its junction with the Watch, the ground being slightly covered with snow, of which a considerable fall took place on the following day.

The nest of the Twite, which is neatly constructed of roots, dry grass, and moss, lined with wool, hair, and feathers, is generally placed amongst heather or grass. The eggs, from four to six in number, are very like those of the Linnet.

THE BULLFINCH.

BULLY, ALP, POPE, NOPE, TONY-HOOP, RED-HOOP, COAL-HOOD, OR COALLY-HOOD.

Pyrrhula europæa.

The Bullfinch.

The Blackbird whistles from the thorny brake;
The mellow Bullfinch answers from the grove.

THOMSON, *Spring.*

THIS handsome bird is a permanent resident in the county,[1] and is more frequently seen in our woods, gardens, and shrubberies during winter and spring, when the trees are bare of foliage, than in summer and autumn, when the leaves usually conceal it from view, although its call-note may be heard.

It appears to be more numerous in certain districts than in others, and it is now prevalent in Lauderdale, where it is said not to have been found thirty years ago.[2] Amongst the favourite haunts of the Bullfinch may be mentioned the woods about Paxton, Wedderburn, Burnhouses, Spottiswoode, Westruther, Legerwood, and Earlston. At Paxton it is plentiful, its nest being found every year in the plantations near the side of the Tweed there.

---

[1] The Bullfinch has been occasionally observed on migration at the Isle of May. *Spring* 1881.—14th April, one seen. *Spring* 1882.—Several seen. *Spring* 1883.—Several seen. There is no record of it being seen at the Farne Islands.—*Reports on the Migration of Birds,* 1879-86.

[2] *Hist. Ber. Nat. Club,* vol. viii. p. 144.

THE BULLFINCH.

The Bullfinch is of rather retiring habits, keeping mostly in the woods, at a distance from houses. In spring it often visits gardens, where it does much damage by destroying the buds of fruit trees and bushes, and is on this account much disliked by gardeners.

It is greatly prized by bird-fanciers in the county, who take it by means of a call-bird and trap-cage, or with bird-lime. It has a low warbling song, which is not heard at any distance.

The food consists chiefly of the seeds of various weeds, such as those of the dock, groundsel, thistle, chickweed, and others, also of haws, elder-berries, and the hips of the dog-rose. I have observed it frequently in my garden in summer, feeding on the seeds of the early-flowering forget-me-not (*Myosotis dissitiflora*). The nest, which is generally placed in a low bush or tree, such as a young spruce-fir, is easily distinguished from that of any other small bird, for it is wholly composed of twigs and rootlets. The eggs are from four to six in number, and are greenish blue, spotted, and occasionally streaked, generally towards the larger end, with purplish brown.

THE PINE GROSBEAK.

THE PINE BULLFINCH, GREATER BULLFINCH.

Pyrrhula enucleator.

> *Who for the worthless bird of brighter plumes
> Would change the meanest warbler of my grove?*
>
> SHENSTONE, *Elegies.*

THERE is a record of the occurrence of this very rare bird in the parish of Eccles, Berwickshire, by Robert Dundas Thomson, M.D., F.R.S., who drew up the revised Report on that parish for the *New Statistical Account of Scotland* in May 1834. Dr. Thomson was a Member of the Royal Societies of London and Edinburgh, and was one of the nine original Members of the Berwickshire Naturalists' Club at its institution on the 22nd of September 1831.[1]

Mr. Seebohm says that the Pine Grosbeak is a circumpolar bird, breeding in the forests at or near the Arctic Circle, and that it inhabits pine-woods and feeds upon the buds of various forest trees, the seeds of fir cones, and the berries of various shrubs, especially those of southernwood.

---

[1] He was born in the Old Manse of Eccles, in 1811.

THE CROSSBILL.

SHELL-APPLE, COMMON CROSSBILL.

Loxia curvirostra.

> *And that bird is called the Crossbill;*
> *Covered all with blood so clear,*
> *In the groves of pine it singeth*
> *Songs like legends strange to hear.*
>   LONGFELLOW, *The Legend of the Crossbill.*

THE peculiar call-note of the Crossbill while passing high overhead in the air, its brightly coloured plumage, and the singular attitudes which it assumes while feeding, all render it an object of attraction. There are, however, very few opportunities of observing this bird in Berwickshire, for it seldom visits the county. Mr. Hardy has informed me that in a letter to him, dated the 9th of May 1857, Mr. W. Cairns, late of Cockburnspath, states that, about 1837, a great flock of these birds frequented Dunglass woods for a time. He has also told me that he had heard from the late Mr. Wilson, of Coldingham, that some Crossbills were seen, and two shot, at Whitecross, about 1838; several having been observed at Whitfield and the Press about the same time. According to Mr. Kelly, a small flock visited Lauderdale in the winter of 1850, and remained for some weeks in the old fir wood opposite Thirlestane Castle, Crossbills appearing again in that locality in 1857.[1] Mr. Cowe has informed me that a male was shot at Dowlaw in 1870. Several small flocks were seen at Paxton in September 1873, where they fed on the cones of the spruce fir trees in the grounds of Paxton House.[2]

---

[1] *Hist. Ber. Nat. Club*, vol. vii. p. 303.
[2] A few birds of this species appeared at Paxton, and also at Drakemire Wood, in the second week of August 1888.

THE CORN BUNTING.

COMMON BUNTING, BUNTING LARK.

Emberiza miliaria.

𝕿𝖍𝖊 𝕮𝖔𝖗𝖓 𝕭𝖚𝖓𝖙𝖎𝖓𝖌, 𝕿𝖍𝖊 𝕮𝖔𝖗𝖓 𝕭𝖚𝖓𝖙𝖑𝖎𝖓𝖌.

> The Craws and Rabbits unco lean,
> By middens and by braes are seen;
> Poor Robin comes to seek a frien'
> To our cot at morn;
> And Buntings in barnyards convene,
> To pick the corn.
>
> DR. HENDERSON, *Winter Rhymes.*

THIS bird appears to have certain favourite localities in the county which it frequents, whilst in other districts, such as the neighbourhood of Paxton, it is very seldom seen. It also seems sometimes, without any apparent cause, to forsake places where formerly it was common. As an instance of this peculiarity, Mr. Charles Watson has informed me that, until about twenty years ago, it was plentiful on the Stonymoor, near Duns, but is now rarely or never seen in that neighbourhood; where, in his boyhood, he was never at a loss for a Corn Bunting's wing when trout flies required to be dressed for the Whitadder. The same habit on the part of the bird has been noticed in the vicinity of Lauder. It occurs in the neighbourhood of Gordon in spring,[1] and is also seen about Cockburnspath,

---

[1] *Hist. Ber. Nat. Club.* vol. ix. p. 230.

Oldcambus, and Penmanshiel, where Mr. Hardy says it may be noticed sitting on dykes singing during summer, and on the tops of bur-thistles in corn and hay fields.[1] Dr. Stuart, writing in 1885, mentions that it is constantly seen about Harelaw,[2] near Chirnside, and on the 4th of January 1886 he records it as observed at Oldcastles.[3] On the 20th of February 1887 I saw a flock of about forty sitting on an ash-tree at Coldlands, near the village of Auchencraw, and heard some of them singing.

The Corn Bunting is found in Berwickshire at all seasons of the year, but it appears to be more plentiful in autumn than at any other time. This may probably be caused by our local birds receiving additions to their numbers at that period, in the shape of migrants from the north on their way southwards for the winter. It has not been often found on migration at the Isle of May,[4] and apparently has never been noticed at the Farne Islands, while it seems to be a local migrant in Berwickshire, removing from exposed to more sheltered localities during winter.

A favourite perch for this heavy-looking bird is the top of a high hedge, or the copestone of a wall; and here in spring and summer it may be seen sitting at its ease in a state of apparent self-contentment, uttering at short intervals its monotonous, but, nevertheless, somewhat pleasing song, which Mr. Hardy has happily likened to the words, "Itzer-teesi-weesi-wees." It lives in pairs in spring and summer, but congregates in flocks in autumn and winter. During severe weather it sometimes visits stackyards in company with Chaffinches and other birds. The

---

[1] Mr. Hardy's *MS. Notes*.
[2] *Hist. Ber. Nat. Club*, vol. xi. p. 239.   [3] *Ibid.* vol. xi. p. 563.
[4] One was seen at Isle of May on 19th Dec. 1882. This is the first recorded instance of its appearance there. One on 12th Feb. 1883; one on 6th April 1883; and two on 31st March 1884.—*Reports on the Migration of Birds*, 1879-86.

Bunting feeds on seeds and grain of various kinds, and also upon insects.

Its nest, which is composed of grass and fibrous roots lined with a little hair, is usually placed on the ground, under the shelter of a bush, or amongst coarse herbage. The eggs, from four to six in number, are purplish white, streaked and blotched with deep brown.

THE YELLOW BUNTING.

YELLOW HAMMER,[1] YELLOW YELDRING,[2] YELLOW YOLDRING, YELLOW YOWLEY, YELLOW YITE, YITE, YOLKRING, YELDROCK, SKITE, DEVIL'S BIRD, GOLDIE.

Emberiza citrinella.

𝔗𝔥𝔢 𝔜𝔢𝔩𝔩𝔬𝔴 𝔜𝔦𝔱𝔢, 𝔗𝔥𝔢 𝔜𝔢𝔩𝔩𝔬𝔴 𝔜𝔬𝔯𝔩𝔦𝔫,[3] 𝔗𝔥𝔢 𝔜𝔢𝔩𝔩𝔦𝔢 𝔜𝔦𝔯𝔩𝔦𝔫, 𝔗𝔥𝔢 𝔜𝔢𝔩𝔩𝔬𝔴 𝔜𝔞𝔪𝔪𝔢𝔯.

Even in a bird the simplest notes have charms
For me: I even love the Yellow Hammer's song.
When earliest buds begin to bulge, his note
Simple, reiterated oft, is heard
On leafless briar or half-grown hedgerow tree;
Nor does he cease his note till autumn's leaves
Fall fluttering round his golden head so bright.
Fair plumaged bird! cursed by the causeless hate
Of every schoolboy, still by me thy lot
Was pitied! never did I tear thy nest:
I loved thee, pretty bird! for 'twas thy nest
Which first, unhelped by older eyes, I found.

GRAHAME, *Birds of Scotland.*

THE Yellow Hammer is plentiful all over the county, and is a permanent resident.[4] The male with his bright yellow head

---

[1] Yellow Hammer signifies Yellow Bunting—the word "Hammer" being derived from the same source as the German word "Ammer," which means a Bunting.

[2] While weary *Yeldrins* seems to wail
Their little nestlings torn.
TANNAHILL, *Songs.*

[3] Sweet waves the green broom where I saw it of old,
And the *Yorlin* is singing beside the sheep-fold.
DR. HENDERSON, *MS. Poems.*

[4] The Yellow Bunting has not been observed in any numbers on migration at the Isle of May, a few only being noticed in the autumns of 1880, 1882, and 1883, and in the springs of 1883, 1884, and 1885. There is apparently no record of its appearance on migration at the Farne Islands in the *Reports on the Migration of Birds,* 1879-85.

is a beautiful bird, and is seen sitting on low trees or hedgerows along our roadsides in spring and summer, uttering his somewhat mournful notes, which have been likened to—"A little bit o' bread, and no che—e—se." Its song continues from early spring until autumn,[1] and is frequently heard on hot summer days when all other birds are silent, being one of the pleasing rural sounds of summer, which charm the ear of the lover of nature. It is likewise one of our earliest morning songsters.

Berwickshire boys are said to have handed down from an ancient date to their successors a hatred of the Yellow Hammer, by calling it one of the Deil's birds; and hence they delighted to harry its nest. They repeated the following rhyme—

> Yellow, Yellow Yorlin,
> Drink a drap o' the Deil's blude
> Ilka Monday morning;[2]

and believed that the devil, crouching in the form of a toad, sat upon its eggs, hatched them, and fed the young—

> Quarter puddock, quarter taid,
> Half a Yellow Yourlie.

---

[1] I heard a Yellow Hammer singing at Paxton in 1888, as early as the 22nd of January. Mr. Hardy has favoured me with the following notes on this bird:—"Penmanshiel and Oldcambus, 18th Feb. 1838.—Singing for first time, although ground covered with snow and hard frost—the sunshine awoke it. 10th Feb. 1872.—Beginning to sing. 17th July 1843.—Sings first, and saluted us about 3 A.M. 23rd July 1868.—Still in song."

[2] Jamieson in his *Scottish Dictionary* says:—"The superstition of the country has rendered it a very common belief among the illiterate and children that this bird (the Yeldring) somehow or other receives a drop of the devil's blood every May morning." Children hang by the neck all the Yellow Hammers they can lay hold of. They often take the bare *gorbals* or unfledged young of this bird and suspend them by a thread tied round the neck to one end of a cross-beam; they then suddenly strike the other end and drive the poor bird into the air. This operation is called *Spangie hewit*. In other parts of Scotland this devoted bird's communications with the devil are believed to be far more frequent, for it is said to receive three drops of his blood every morning. Mr. Hardy says that he heard this rhyme from a gentleman who passed his youth in the hills between Berwick and Roxburgh shires, where it may still linger.—*Hist. Ber. Nat. Club.* vol. i. p. 219.

THE YELLOW BUNTING.

In revenge for the persecution which it received the Yellow Hammer cursed its persecutors in its song, which was thus interpreted—

> Deil, Deil, tak ye;
> For me to big a bonny nest,
> An' you to take it frae me.

Mr. Hardy writes that this bird was also popularly believed to be difficult to shoot, and if shot the devil would take the offender.

I have not heard that the boys of the present day believe in the folk-lore of their fathers with regard to this bird.

In winter the Yellow Hammer associates with flocks of Chaffinches, Greenfinches, and other birds in our stubble fields, to pick up seeds and shed corn. During severe weather, when snow is on the ground, it may be seen in the farm-yards, on the tops and under the eaves of stacks, and about barn-doors and cattle-sheds, picking up what it can get in the way of food. In summer it lives chiefly upon insects.

The Yellow Hammer is a somewhat late breeder. The nest, which is composed of dry grass with some moss, and is lined with hair, is generally placed near the ground, under the shelter of a bush or by the side of a hedge or ditch. I have, however, seen it among strawberry plants, at the foot of a garden wall at Paxton. It is a favourite nest with boys, being easy to find, and is often the first bird's nest which they discover. The eggs are four or five in number, dull white, streaked and spotted with purplish brown.

THE REED-BUNTING.

BLACK-HEADED BUNTING, RING-BUNTING, RING-BIRD, REED-SPARROW, WATER-SPARROW, BLACK-BONNET, BLACK-CAP.

Emberiza schœniclus.

The Coal Hood, The Coal Hooden.

> "*Heichow! wae's me, that I sude hae lived to see the day! That ever I sude hae lived to see the Cole-hood take the Laverock's place, and the Stanechel and the Merlin chattering frae the Cushat's nest.*"
> JAMES HOGG, *Brownie of Bodsbeck.*

THE Reed-Bunting is found thinly scattered over the county, in bogs and marshy places

> Where meadow-blooms, red, drooping hang.
> O'er stream and pool, and bog reeds lang
> Wave slowly, and bull-rushes strang
> Shade Kelpie's Hole.[1]

Gordon Bog, where it nests, is one of its favourite haunts; and there during summer the male, with his black head and pretty white collar, may be seen perched on a willow, alder, or other bush, occasionally enlivening the waste with his short song. In former times, before the advance of agriculture had converted so many of the numerous marshes and swamps with which the county then abounded into cultivated fields, this species would doubtless be much more plentiful than it is now.

---

[1] Dr. Henderson's *MS. Poems.*

Although the Reed-Bunting is found in Berwickshire throughout the year, yet it appears to be a partial migrant, for fewer are seen in the county during winter than in summer. It does not seem to have been observed on migration at the Farne Islands or the Isle of May, with the exception of a single specimen which was noticed at the latter place on the 1st of March 1882, but there are records of it having been seen on migration on the Lincolnshire and Norfolk coasts in autumn.[1] It has been observed in winter among the reeds by the side of the Tweed and Whitadder near Paxton; and my friend, Mr. Hardy, in his notes on birds in the neighbourhood of Penmanshiel and Oldcambus, states that it is occasionally found on the moors of Coldingham and Redheugh at that season; also, that it resorts to stacks of corn in the fields at Penmanshiel for food during hard weather.[2] He mentions Lumsden Moss as a summer resort.

A male and female were caught with bird-lime on some whins near Whiteburn by the gardener at Abbey St. Bathans in February 1886. They had been feeding on the ground from which a stack of corn had been removed. The Rev. George Cook, of Longformacus, informs me that this species is often seen about Whitchester. Mr. Lockie, Spottiswoode, says that it is becoming scarcer every year in the Earlston district. Before the extensive morass of Billie Mire was drained, the great beds of reeds which abounded there would form a favourite resort of this bird; and, even yet, it frequents the neighbourhood of the site of the Mire, for Dr. Stuart records seeing several Reed-Buntings in February 1885 at Billie Brae, where, he says, they nest. He likewise mentions that three were

---

[1] *Reports on the Migration of Birds*, 1879-86.
[2] I saw several Reed-Buntings haunting some stacks by the side of the public road near Whiterigg in December 1887.

observed in the stackyard at Foulden West Mains about the same time.[1]

The food of the Reed-Bunting in summer consists chiefly of insects, and in winter it eats the seeds of grass and other plants, as well as grain.

The nest, which is composed of dry bents lined with hair, is generally built close to the ground, among rank vegetation or rushes,[2] but I have sometimes found it in low willow bushes in a saugh bog.[3] The eggs, which are from four to six in number, are usually of a purple brown colour, spotted or blotched, and streaked with dark brown or black.

---

[1] *Hist. Ber. Nat. Club*, vol. xi. p. 236.

[2] Mr. Compton-Lundie, of Spital, showed me a nest with five young amongst rushes, near the lily pond at Spital House, on 11th June 1888.

[3] When a boy I used to search for the Coal-Hooden's nest in a saugh bog near the Uuder Kilns at Salton, East-Lothian, and occasionally found it there. Its beautiful coffee-coloured eggs were considered "rare."

THE SNOW-BUNTING.

SNOW-FOWL, SNOW-FLECK, SNOW-FLAKE, TAWNY BUNTING,
MOUNTAIN BUNTING, GREAT PIED MOUNTAIN FINCH.

Plectrophanes nivalis.

The Snaw-Fleck.

But O the Snaw-Fleck!
The bonny, bonny Snaw-Fleck,
She is the bird for me, O.
JAMES HOGG.

THE local name of this interesting and pretty bird is very appropriate; for in the dark lowering weather which usually precedes a snow-storm in winter, the white marks on the parti-coloured plumage of a flock of Snow-Buntings on the wing, coupled with their wavering flight, somewhat resemble flakes of falling snow.

The Snow-Bunting is an autumn and winter visitor from the north, arriving in Berwickshire from about the third week in October to the middle of November, and leaving us again for northern regions towards the end of March.[1] It is seen, while on migration, passing the lighthouses on the east coasts of England and Scotland in vast numbers in autumn and spring.[2]

---

[1] In Autumn it has been observed at Penmanshiel as early as 2nd October (1883).—*Hist. Ber. Nat. Club*, vol. x. p. 569. In spring it has been noticed at Oldcambus as late as 29th March (1882).—*Ibid.* vol. x. p. 557.

[2] Snow-Buntings were observed on migration in Autumn 1879.—At Farnes, 5th Sep., one at the Longstone (Mr. Cordeaux says this is the earliest example ever recorded in England). *Autumn* 1880.—Isle of May, 27th Oct. to 13th Nov.; Farnes, 28th Oct. to 20th Dec. *Spring* 1881.—Isle of May, 9th April. *Autumn*

On its first arrival in flocks, which is often the precursor of stormy weather,[1] it frequents the lea and stubble fields in the neighbourhood of the sea-coast, and is shortly afterwards observed on the uplands of the Lammermuirs. I saw a large flock near Longformacus in November 1874, and Mr. Lockie, Gateside School, Spottiswoode, states that it is numerous in winter on Corsbie Muir, and likewise in the Westruther, Spottiswoode, and Earlston districts. Mr. Tilly, Lauder, writes that he has frequently noticed flocks on Trabroun Farm in Lauderdale, at the same season, where their arrival is regarded as the sign of an approaching heavy snow-storm.[1] Mr. John Wilson, of Welnage, Duns, has informed me that he has often seen great flocks on Edington Mains Farm, in stormy weather in winter. Mr. Hardy, who has given many interesting notes in the *History of the Berwickshire Naturalists' Club* on the migration of this bird, mentions Penmanshiel, Oldcambus, Bowshiel, Dowlaw, Lamberton, and Fairneyside, as some of its favourite resorts in the vicinity of the sea-coast.[2]

In winter the Snow-Bunting assembles in large parties, which are continually itting about their feeding-grounds, uttering their tinkling call-notes, and often wheeling rapidly

---

1881.—Isle of May, 24th Sep. to 10th Dec.; Farnes, 6th Oct. to end of year. *Spring* 1882.—Isle of May, 13th Feb. to 15th April; Farnes, in March. *Autumn* 1882.—Isle of May, 19th Sep. *Spring* 1883.—Isle of May, 23rd March; Farnes, 7th March. *Autumn* 1883.—Isle of May, 21st Sep. to 18th Jan. 1884. *Spring* 1884.—Farnes, 11th March to 7th June; one on rocks at latter date. *Autumn* 1884.—Isle of May, in Oct. *Spring* 1885.—Isle of May, 7th Feb.; Farnes, 4th March. *Autumn* 1885.—Isle of May, in Sep., Oct., and Nov.—*Reports on Migration of Birds*, 1879-85.

[1] Mr. Cordeaux, in his *Report on the Migration of Birds* on the East Coast of England in the Autumn of 1880, says:—"The immigration of Snow-Buntings on the north-easterly coasts in four distinct rushes, namely, at the end of October, November, December, and January 1881, has been attended with corresponding depressions of the barometer, and outbursts of arctic weather." Mr. Harvie-Brown, in his account of migration on the east coast of Scotland in the autumn of 1883 (Isle of May), states that "Snow-Buntings often appear with gales of snow and sleet;" and Mr. Hardy, in his *MS. Notes*, writes—"Mr. Hood, Townhead, always prophesied snow on the arrival of the Snow-Flakes."

[2] See vols. vii., viii., ix., and x. of the *Hist. Ber. Nat. Club.*

round before alighting on the ground, where they run about like wagtails. The flocks are very erratic in their movements, changing about from place to place, according to the the weather. They feed on various grass seeds, which they find on the leas and moors, and likewise on the waste corn in the stubble fields; but when severe weather sets in, and

> the stormy north sends driving forth
> The blinding sleet and snaw,
>
> BURNS, *Winter*.

they are sometimes driven to visit stackyards in the vicinity of their haunts, to pick up what they can get there in the way of food.[1]

---

[1] Mr. Kelly records that on 14th November 1880 he saw hundreds of Snow-Buntings feeding on the corn stacks on Lauder Common; also that, where hill sheep are fed on hay, they resort in great flocks to feed on the seed. He adds that they were plentiful at Longcroft. Mr. Hardy states that in severe winters, particularly in drifty days, he has seen them frequenting Penmanshiel stackyard. —*Hist. Ber. Nat. Club*, vol. ix. pp. 404-5.

THE STARLING.

STARE, COMMON STARLING.

Sturnus vulgaris.

𝕿𝖍𝖊 𝕾𝖙𝖎𝖗𝖑𝖎𝖓.

> *Syne, at the midis of the meit, in come the menstrallis.*
> *The Maviss and the Merle singis,*
> *Osillis and Stirlingis.*
> HOLLAND, *The Houlat,* c. 1453.
>
> *The garruling of the Stirlene gart the Sparrow cheip.*
> SIR DAVID LINDSAY, *Complaynt of Scotland.*

MR. JOHN WILSON, late of Edington Mains, who has lived in Berwickshire for the last seventy-seven years, has informed me that the Starling was a rare bird in the county about the beginning of this century, and that it was then met with so seldom, that he was sixteen years old before he saw one. In those days, Mr. Wilson says, a Starling's nest was considered by boys at school to be a great prize, and was spoken about by them for a long time after its discovery. Dr. Stuart, Chirnside, has mentioned to me that when he came to reside in Berwickshire, upwards of thirty-five years ago, the Starling was rather scarce.

It has now become so plentiful that it abounds in every part of the county, and in the autumn months flocks of many thousands roost together in some of the extensive shrubberies which surround the houses of the landed proprietors, the numbers of the birds in some cases being so vast as to break down the branches upon which they alight.

Before the destruction of the tall evergreens at Paxton House and Ayton Castle, by the severe winters of 1878-1881, they were frequented by immense flocks of Starlings at night, and the shrubberies about Mordington House were likewise a favourite resort in the autumn of the former year. Mr. Loney, Marchmont, has been so good as to send me a very interesting account of the roosting of the Starling in the evergreens there, in the autumn of 1885 and 1886, in which he says:—"A belt of common laurel, with an uncut beech hedge fifteen feet high, was their favourite place for the night. It was no uncommon sight to see twelve or more birds on a beech bough about two feet long, and when they settled down, as they usually did, in large masses, the hedge seemed as if it were lowered two feet by their weight. The chattering which they made was very great; sometimes they would all stop at once, and remain silent for a minute or two, and then, on what appeared to be a preconcerted signal, they would resume their chorus, which, on the hands being clapped, would again instantly cease. As they were becoming a nuisance, a man was employed with a gun to fire blank cartridges at them, to try to drive them away; but they got so accustomed to this, that they would rise on the gun being fired, and fly round and round several times in a body, and again alight in their old quarters. They left about the end of October, only a few birds remaining through the winter. In the autumn of 1886 they arrived on the 20th of August, nearly a month earlier than in 1885, and in greatly increased numbers. Not only did they take possession of their old roosting-place, but also of the laurels to the west of Marchmont House. Here a number of pheasants roosted, but the advent of the Starlings did not seem to disturb them in the least. Very early in the mornings the whole of the Starlings commence their chattering almost simultaneously; this continues for a minute or two, when they take wing in companies, and fly off to their feeding-grounds, generally visiting in the first place the grass parks, where sheep have been lying over night."

Mr. Loney has likewise informed me that, in the summer of 1887, numbers of Starlings continued to frequent their usual roosting-places at Marchmont during the nesting season, and as they did not appear to be breeding, several specimens were shot for dissection, which all proved to be males. Colonel Brown of Longformacus tells me that, in the winter of 1886-87, the dovecot, which stands near the mansion-house there, was taken possession of by these birds, which drove out the pigeons, and occupied the dovecot all the winter, many hundreds roosting in it at night. Starlings had not previously been known to stay over the winter at Longformacus.

Although this species remains in the county throughout the year, it is a local migrant; for, when severe weather sets in, with hard frost and snow, it leaves the more inland parts, and frequents the neighbourhood of the seacoast. It may sometimes be seen in such weather, about farm-yards, picking up waste corn and seeds about the stacks and barn doors.[1] At all seasons the ground upon which sheep are feeding seems to be a favourite resort, and it may be often seen perched on their backs searching for the insects which infest their wool. It is, undoubtedly, one of our most useful birds, and the quantity of injurious insects and grubs which it destroys is immense.

The Starling is seen in great flocks on migration at

---

[1] On 31st December 1874, during the severe weather which then prevailed, I saw several Starlings feeding about the corn-barn door at Nabdean. Mr. Hardy mentions (*Hist. Ber. Nat. Club*, vol. ix. p. 125), that in the protracted snowstorm of 1878-79, vast numbers of Starlings frequented the stackyards at Bowshiel and Oldcambus.

the Isle of May, and sometimes at the Farne Islands, in autumn, and occasionally in spring.[1] It is probable that our Berwickshire birds receive additions to their numbers from these in early autumn, and help to increase the migratory flocks later in the season.

Even before many of our early spring flowers come into bloom, the mellow whistle of the Starling on its return to its accustomed nesting-place is heard, and it tells us that the winter is past and gone, and that the time of primroses is at hand—

> Of all the birds whose tuneful throats
> Do welcome in the verdant spring,
> I prefer the Steerling's notes,
> And think she does most sweetly sing.
>
> ALLAN RAMSAY.

It builds in old towers such as those of Corsbie, Whitslaid, Evelaw, and Dryburgh Abbey; also in rocks overhanging streams, as on the Tweed, Whitadder, and Leader; and on the sea-coast, as at Siccar; likewise in holes in trees and walls; under the eaves and about the chimneys of houses; and in various other situations. The nest is generally composed of dry grass and straw, with roots and twigs, and the eggs, which are from four to seven in number, are of a pale blue colour, without any markings.

Young Starlings are sometimes taken from the nest by bird-fanciers in the county to be kept as cage-birds, and taught to speak. This seems to have been done in Scotland as long ago as the time of Sir David Lindsay, for, writing in

---

[1] The Starling was seen on migration in *Autumn* 1879.—Farnes, 22nd Sep. *Autumn* 1880.—Isle of May, 5th July, 12th Sep.; Farnes, 15th Oct., 13th and 19th Nov. *Autumn* 1881.—Isle of May, 21st Oct. *Autumn* 1882.—Isle of May, 1st July, 7th August, 25th Dec. *Autumn* 1883.—Isle of May, 1st Nov., 31st Dec. *Spring* 1884.—Isle of May, 15th Feb. *Autumn* 1884.—Isle of May, 26th Dec. *Spring* 1885.—Isle of May, 9th Feb., 11th March. *Autumn* 1885.—Isle of May, 17th Oct. to Jan. 1886.

1592, and referring to some of the religious exercises of the time, he says—

> . . . . . . I thiuk ane greit derisioun
> To heir thir nunnis and sisteris nycht and day,
> Singand and sayand psalmis and orisoun;
> Nocht understanding quhat thay sing nor say.
> Bot, like ane Stirling, and ane popingay,
> Quhilk leirnit ar to speik be lang usaye.

The plumage of the Starling seldom varies. A white specimen was, however, got some years ago near Birgham by Mr. Waddell.

THE CHOUGH.

CORNISH CHOUGH, CORNISH DAW, CORNWALL KAE, CHAUK DAW, KILLIGREW, RED-LEGGED CROW, MARKET JEW CROW.

Pyrrhocorax graculus.

The Red-nebbed Crow.

Come on, sir; here's the place; stand still; how fearful
And dizzy 'tis, to cast one's eyes so low!
The Crows and Choughs, that wing the midway air,
Show scarce so gross as beetles: half-way down
Hangs one that gathers samphire,—dreadful trade!
Methinks he seems no bigger than his head:
The fishermen, that walk upon the beach,
Appear like mice.
            SHAKESPEARE, *King Lear.*

THE late Dr. George Johnston, of Berwick-on-Tweed, in his address to the Berwickshire Naturalists' Club, at its first Anniversary Meeting, on the 19th of September 1832, says, with regard to the visit of the Club to St. Abb's Head in the previous July:—" I must not leave this majestic coast without mention of another of its feathered tenants, the Cornish Chough, which indeed was not seen by us on this occasion, but is certainly ascertained to breed in the rocks between St. Abb's Head and Fast Castle.[1] This fact, distinctly

---

[1] Mr. Hancock, in his *Birds of Northumberland*, says:—" With regard to the Chough, a specimen in my collection was presented to me by the late Mr. George Johnston, of Berwick-on-Tweed; it was shot at Redheugh, near the place where it was breeding." Mr. Allan, Redheugh, says that Dr. Johnston's brother was tenant of that farm from 1823 to 1842.

mentioned by Bishop Lesley in his history *De Origine Scotorum*, published about 300 years ago,[1] has been overlooked or disregarded by naturalists, who have considered the bird peculiar to the western shores of Britain; and it is to the Rev. A. Baird [2] that we are indebted for the confirmation of the accuracy of the Bishop's information; and, of course, for showing that the limits usually assigned to the distribution of the Chough in this country are erroneous."[3] Selby, writing on the *Ornithology of Berwickshire* in 1841, mentions that it "finds a congenial retreat in the precipices of St. Abb's Head and adjoining coast," and that "here it is not uncommon, but being a bird of wary habit, it is very difficult to approach within gunshot, and specimens are not easily obtained."[4] In 1846, Mr. Hardy noted that a pair were then at Fast Castle, and that the young used formerly to be climbed for, and taken out of the nests, to be tamed.[5] Mr. Archibald Hepburn, in a paper "On some of the Mammalia and Birds found at St. Abb's Head" in 1851, says that "the Chough or Red-legged Crow is now extinct, except a solitary pair," which, according to his information, "seldom strayed far from Fast Castle, a few miles to the eastward of the Head."[6] In 1855, Mr. Wilson, Coldingham, wrote that the "Red-legged Crow or Kay built formerly at Biter-cove and Thrummycarr Heugh, but is now extinct in this neighbourhood."[7] Dr. Turnbull, in his *Birds of East-Lothian*, written in 1866, alluding to this bird at St. Abb's Head, says:—"It is still

---

[1] Published in 1578, ed. 1675, p. 17.
[2] Assistant minister of the united Parishes of Cockburnspath and Oldcambus. He wrote the Report on these parishes in the *New Statistical Account of Scotland* in 1834. He was one of the original members of the Berwickshire Naturalists' Club.
[3] *Hist. Ber. Nat. Club*, vol. i. p. 6.   [4] *Ibid.* vol. i. p. 253.
[5] From *MS. Notes* by Mr. Hardy.
[6] *Hist. Ber. Nat. Club*, vol. iii. p. 72.
[7] From *MS. Notes* by Mr. Hardy.

seen there, but is supposed to have dwindled in that locality to a single pair."[1] Mr. Robert Cowe, Oldcastles, near Chirnside, who is a native of the parish of Coldingham, and lived for many years in the vicinity of Fast Castle and St. Abb's Head during the early part of this century, has informed me that between 1820 and 1830, when he was a boy attending school in the village of Coldingham, he often saw the Red-nebbed Crow about the rocks of the sea-coast immediately to the west of Petticowick, and that it built in a steep precipice there; also that his school-fellows sometimes took the young ones to be kept and tamed like Jackdaws.

The Chough appears to have become extinct about St. Abb's Head and Fast Castle between 1846 and 1855, and to have remained so; for had this not been the case, the bird would surely have been seen after the last-mentioned date by fishermen and others; including persons well acquainted with its appearance, who have been long resident in the neighbourhood of St. Abb's Head and Fast Castle, and who have from time to time passed by its former haunts at all hours of the day, from early dawn until dark, for the last thirty years. Notwithstanding numerous inquiries, I have not been able to find any person who has seen it on the Berwickshire coast within the period mentioned. The great increase in the number of Jackdaws round the coast has probably been the cause of its disappearance.

This species is rather less in size than the Rook, to which it bears a resemblance, but it is more elegant in form. When near, it is at once distinguished by its red bill and feet.

---

[1] *Birds of East-Lothian*, p. 18.

THE JAY.

JAY PIE, JAY PIET.

Garrulus glandularius.

The Jay Piet.

*I herde the Jay and the Throstell,
The Mavis meynd[1] in hir song,
The Wodewale farde as a bell,
That the wode about me rung.*
     TRUE THOMAS.[2]

THE Jay was in former times a well-known bird in the wooded districts of the county, but now it is so rare that it is very seldom seen, having been completely extirpated by gamekeepers many years ago, on account of its egg-sucking propensities. The Rev. Andrew Baird, in his account of the united parishes of Cockburnspath and Oldcambus, written in 1834, mentions that it then built in considerable numbers in Penmanshiel Wood. Mr. Hardy states that, some time before that period, its numbers had been greatly reduced by trapping, and it was rooted out there at a somewhat later date. In 1837, it was still frequenting the neighbourhood; for in a Naturalist's Calendar, kept at Penmanshiel by Mr. Hardy's brother, two Jays are noted as having been seen on the 2nd of May that year. About the same time it used to frequent Dunglass Dean, and Mr. Hardy remembers that in his youth he often saw it nailed up by the head to game-

---

[1] To lament.
[2] Jamieson's *Popular Ballads and Songs*, Edin. 1806, vol. ii. p. 11.

keepers' vermin-rails in the vicinity. Writing in 1872, he says :—" There have not been any Jays there for well-nigh thirty years." [1] Mr. William Patterson has informed me that when he was a boy at Swinton School, about fifty years ago, these birds were plentiful in Duns Castle woods, and there were also some about Marchmont. Mr. W. Duns lately showed me a stuffed specimen which had been killed near Duns Castle about 1856, and stated that John Fairlie, who was then gamekeeper there, used frequently to trap and shoot Jays at that period in the woods. Mr. Turnbull, of Abbey St. Bathans, has told me that they used at one time to frequent the woods about The Retreat; and Mr. Craw, Foulden West Mains, has mentioned that one was killed near Dye Cottage about thirty-five years ago, and is still preserved by a shepherd in that neighbourhood.

Although I have not been so fortunate as to meet with the Jay in Berwickshire, I have seen great numbers near Remorantin in France, where they used to frequent the orchards and woods in small flocks, and were very shy and wary. It is to be regretted that this handsome bird has been totally exterminated in the county.

[1] *Hist. Ber. Nat. Club*, vol. vii. p. 514.

THE MAGPIE.

PIE, PIET, PYET, PYOT, PYE, PIANET, MADGE, MAG.

Pica rustica.

The Piet, The Pyet, The Pyot.

The Pyet with hir pairtie cot,
Fenyeis to sing the Nichtingalis not;
Bot scho can nevir the corchet cleif,[1]
For harshnes of hir carlich[2] *throt.*

W. DUNBAR, ABOUT 1490.

DURING the first quarter of this century, and until game began to be generally preserved, this interesting bird was plentiful in Berwickshire, but in many districts of the county it is now seldom seen, having been almost extirpated by gamekeepers on account of its partiality for the eggs and young of Pheasants and Partridges. Mr. Wilson, late of Edington Mains, says that in his boyhood, between 1810 and 1820, Piets were always to be seen in considerable numbers in the neighbourhood of that farm; and this continued until about 1840, when, along with Hooded Crows, polecats, and other enemies to game, they were nearly exterminated. According to Mr. Duns, Duns, they were numerous about Broomhouse in 1828, and he relates that when he was working there at that time as an apprentice mason, at the erection of the stone pillar which stands on the edge of the

---

[1] "Divide a crotchet," a term of music.—Sibbald, *Chron. Scot. Poet.* i. p. 319.
[2] Coarse, vulgar.

bank to the east of the mansion-house, General Home, who disliked Piets very much, used to get him to climb the trees on which they built, to harry their nests. Mr. Hardy mentions that Penmanshiel Wood was a favourite resort of this bird until 1843, when it was extirpated by the game-watchers. It is still, however, frequently seen in the parishes of Legerwood,[1] Westruther, Lauder,[2] Abbey St. Bathans,[3] and Coldingham,[4] where there are extensive woods, and occasionally about Mertoun and Cowdenknowes, but it is also found in other districts. A pair had their nest and reared their young in a small fir wood at Paxton in 1874, and as the young birds were allowed to fly, an occasional specimen was observed in the neighbourhood for some years afterwards.

The Piet used to be considered a bird of omen by the peasantry of the county, and, as the belief still lingers, the following rhyme is occasionally heard in some localities:—

> One's mirth, two's grief,
> Three's a wedding, four's death,
> Five is heaven, six is hell,
> Seven the devil's ain sel'.

Mr. Duns states that an old man named Jamie Dewar, who was a cabinetmaker in Duns about thirty-five years ago, believed so firmly in the truth of the above lines, that he would sometimes turn back when on an important errand if two Piets happened to appear on the road. As another

---

[1] *Hist. Ber. Nat. Club*, vol. ix. p. 242. Mr. John Logan informed me on 22nd Feb. 1887 that Magpies were so plentiful about Legerwood that he had to employ a man to kill them. They bred in the strips of wood on the farm, and were still numerous in 1886 when he left Legerwood.

[2] While driving past Huntingdon, near Lauder, on 3rd March 1885, I saw five or six Piets in a strip of wood by the side of the road.

[3] Mr. Turnbull of Abbey St. Bathans is good enough to give his gamekeeper directions not to kill out all the Piets, and on that account a pair or two may be seen about Bushelhill.

[4] There is a nest every year on the old Scotch fir trees near the millpond at Sunnyside.

instance of the popular superstition with regard to this bird, Mr. Clay, Kerchesters, relates that about thirty years ago, when it was often seen in the neighbourhood of Winfield, where he then lived, and when his eldest son was a little boy, an old nurse called Mary Lorraine, who attended to him, came home one day from a walk and told Mrs. Clay that, as she had seen seven Picts together, she was sure that something very serious was going to happen. About a week afterwards Winfield farm-steading was burned down, and the old nurse thought that the seven Picts had foretold the misfortune. There are several versions of the above rhyme current, in which one Piet is said to be a good, and two a bad omen;[1] but in some districts of France, the popular belief is the reverse, for the saying there runs—" Voir deux pies ou deux corneilles, c'est du bonheur, n'en voir qu'une seule, c'est du malheur."[2]

The following places in Berwickshire have apparently derived their names from having been much frequented by Piets :—Pyatshaw-knowe, a hill (1162 feet), and Pyatshaw Ridge (1250 feet,) above Byrecleugh, in the parish of Longformacus ; Pyotknowes, about a mile south of Marchmont in Fogo parish; Pyatshaw, a wood in Westruther parish, a short distance east of the Dod Mill on the Lauder Road ; and Pyatshaw Burn, which flows into the Brunta Burn in the same neighbourhood.

This species is generally very shy and watchful, and is almost constantly on the move.

> From bough to bough the restless Magpie roves,
> And chatters as she flies.
>
> GISBORNE, *Walks in Forest.*

It will not allow a person to approach within gun-shot, but

---

[1] *The Folk Lore and Provincial Names of British Birds*, by the Rev. Charles Swainson, 1886, pp. 77, 78.
[2] Eugène Rolland, *Faune Populaire de la France*, tom. ii. p. 140.

usually keeps flitting from tree to tree, or along dykes or hedges, when followed, until it at last flies off to a distance. It is, however, very easily poisoned or trapped, and this has undoubtedly led to its extirpation in many districts, where its total destruction is to be regretted by all lovers of birds, for its occasional appearance gives an additional interest to a walk or drive in any locality. Although the food of the Piet varies considerably, it chiefly consists of worms, snails, slugs, and insects of various kinds. As already mentioned, it is very destructive to the eggs and young of winged game ; it also robs small birds' nests, and the eggs of domestic fowls and ducks are greedily devoured when they happen to be laid away from the farm-steading in the bottom of some hedge or covert frequented by the plunderer.

It breeds early in spring, and usually builds its nest in high trees, often choosing the top of an old bushy Scotch fir for the purpose.

> .   .   .   .   .   For skill
> To build his dwelling few can vie
> In talent with the artful Pie :
> On turf-reared platform intermixt
> With clay, and cross-laid sticks betwixt.
> 'Mid hawthorn, fir, or elm tree slung,
> Is piled for the expected young
> A soft and neatly woven home :
> Above of tangled thorns a dome
> Forms a sharp fence the nest about,
> To keep all rash intruders out.
>
> BISHOP MANT.

The eggs, from six to nine in number, are pale bluish green, closely freckled with greenish brown. A farming adage of the olden time in Berwickshire was that "It is not too late to sow bear when the leaves cover the Pyet's nest," which is usually in June. This saying is mentioned by the Rev. Walter Anderson, D.D., in his

Report on the parish of Chirnside, in *The Old Statistical Account of Scotland*, 1795 (vol. xiv. p. 10).

The young are sometimes taken from the nest and taught to speak, and this has apparently given rise to the nickname of "Tale Piet" or "Piet Tongue," which is given by boys to a schoolfellow who is guilty of tale-bearing.

THE JACKDAW.

DAW, JACK, KAE, KAY.

Corvus monedula.

𝕿𝖍𝖊 𝕵𝖆𝖈𝖐, 𝕿𝖍𝖊 𝕵𝖆𝖈𝖐𝖉𝖆𝖜, 𝕿𝖍𝖊 𝕶𝖆𝖞.

> *Deir on deis and thou be dicht,*[1]
> *And syne sits drowpand lyke a da,*
> *Fayn will thay all be of that sicht;*
> *And thay that onlie is thy fa,*
> *They will nocht gruge to lat ye ga.*[2]
>
> JOHN MAITLAND, *Advice to be Blyth*, 1566-70.

THE Jackdaw is now much more plentiful in Berwickshire than it was in former times, and breeds in great numbers in the precipices about St. Abb's Head and all round the coast,[3] as well as in the picturesque ruins of old Dryburgh Abbey, whilst at Corsbie,

> From hollows of the towers on high,
> The gray cap'd daws in saucy legions fly.
>   BLOOMFIELD.

Precipitous cliffs overhanging the rivers and streams of the county are also favourite haunts of this bird. Amongst these resorts may be mentioned the rocks on the Tweed,

---

[1] Mr. Pinkerton says this means, "Though thou be dearly (richly) dressed, and sitting in the place of honour."—Sibbald, *Chron. Scot. Poet.*, vol. iii. p. 318.

[2] The "Advyce to be blyth in bail," from which these lines are quoted, is supposed by Mr. Pinkerton to have been written by "John Maitland, Commendator of Coldinghame, and sone aftir Lord Thirlstane and Chancellor of Scotland."—Sibbald, *Chron. Scot. Poet.* vol. iii. p. 318.

[3] It is supposed that the great increase of the Jackdaw on the sea-coast has led to the extinction of the Chough there.

near Milne Graden; at Edrington Castle, on the Whitadder; below Blackadder House, on the Blackadder; the steep rocky braes about Carolside and Chapel, on the Leader; and the Jackdaw's Craig,[1] on the Dye, near Longformacus. It is very numerous in the neighbourhood of Oldcambus, and my friend Mr. Hardy frequently refers to it in his notes on birds in that district.[2] As an instance of its curiosity, he mentions that on the 1st of April 1863, a ewe dropped two lambs in the old churchyard of St. Helen's, near his house; and on visiting the place to see the lambs, immediately after the event, he found a great number of Jackdaws assembled on the kirk gable and kirkyard dyke, in solemn consultation, as if they could not understand what had occurred, their attention having been attracted by the new "ferlie" while passing on their morning flight.

Jackdaws may be often seen associating with flocks of rooks, and are undoubtedly of service to the farmers by helping to destroy grubs of all kinds, but they sometimes attack young beans as they are springing through the ground, and also newly sown grain; while they steal from corn fields and stacks. Fields where sheep are feeding are favourite resorts of this bird, where it may be frequently observed sitting on their backs, and pecking insects out of the wool. It is much disliked by gamekeepers, for it is very destructive to the eggs, as well as the newly hatched young, of Pheasants and Partridges; pouncing down and seizing the chicks on every opportunity, especially in the morning.

In addition to the above-mentioned situations, the nest is sometimes found in holes in trees and in rabbit burrows. It is chiefly composed of sticks and is lined with wool, dry grass,

---

[1] Colonel Brown, who kindly showed me this haunt on 10th March 1887, said that it is frequented by hundreds of Jackdaws.

[2] *MS. Notes* by Mr. Hardy.

or other soft materials; but the Jackdaw does not seem to confine itself to these comfortable surroundings, as the following account, which appeared in the *Berwickshire News* of the 5th of May 1887, shows:—"While the chimneys of the mansion-house of Mellerstain were being cleaned on Saturday last, a Jackdaw's nest was found in one of the vents. On the nest being taken out, the workmen were greatly surprised to find among the materials composing it, a will belonging to a man named Robertson, who had lived at one time near Mellerstain. Wonderful to relate, it was entire, with the exception of a small bit torn off one of the corners. It was sent off to the relatives, who live near Leitholm. How it came there is a mystery. The thieving propensities of the Jackdaw are well known, but surely it is rare that a 'will' is carried off for the purpose of making a nest."

The eggs, which are from four to six in number, are bluish green, freckled and spotted with ash grey and olive brown. The young are often taken from the nest by boys to be tamed and kept as pets.

THE CARRION CROW.

BLACK CROW, BLACK-NEB, CORBIE CROW, HODDY, HOODIE, GOR-CROW.

Corvus corone.

𝕿𝖍𝖊 𝕮𝖔𝖗𝖇𝖎𝖊 𝕮𝖗𝖆𝖜, 𝕿𝖍𝖊 𝕳𝖔𝖔𝖉𝖎𝖊.

They carried him to the good greenwood,
Where the green pines grow in a row;
And they heard the cry from the branches high,
Of the hungry Carrion Crow.
LEYDEN, Lord Soulis.

Yit by my selfe I find this prouerbe perfyte,
The Blak Craw thinks her awin birdis quhyte.
GAWIN DOUGLAS, Virgil.

THE Carrion Crow, Black Crow, or Corbie Craw, is rather common in Berwickshire, and is found about many of the woods and plantations, where its harsh croaking note generally indicates its presence before it is seen. In autumn and winter it seems to prefer the neighbourhood of the sea-coast. Although it is a permanent resident in the county, it appears to receive additions to its numbers in autumn from migratory flocks, and movements are also observed in spring, at both of which seasons it has been frequently noticed in small scattered bands passing high overhead at Paxton, for several days in succession.[1]

---

[1] The Carrion Crow has been observed on apparent migration at the lighthouses on the Isle of May and the Farne Islands in spring and autumn, viz.— *Autumn* 1881.—Isle of May. *Spring* 1882.—Isle of May, 14th May. *Autumn*

THE CARRION CROW. 211

Some of our leading ornithologists[1] have now come to the conclusion that the Carrion Crow (*Corvus corone*) and the Hooded Crow (*Corvus cornix*) are only different forms of the same species, and the fact of the two interbreeding freely in a wild state,[2] in some parts of Europe and Asia, supports this view. The habits of the Carrion Crow and those of the Hooded Crow, as seen in this county, differ in some respects—the former frequenting inland districts and woods more than the latter, which is usually found about the coast, or on open ground, such as fields where sheep are feeding on turnips. Nearly all the Hooded Crows likewise leave us before summer, while numbers of Carrion Crows remain during that season, and nest in the woods.

This species is a great enemy to game of all kinds, seizing and making off with the young, and searching hedgerows and covers for the eggs of Partridges and Pheasants, which it greedily devours. On this account it is ruthlessly shot and trapped by gamekeepers. It likewise commits depredations amongst poultry by carrying off young chickens, which it generally does early in the morning.[3]

1882.—Farnes, with Grey Crows in Oct., Nov., and Dec. *Spring* 1883.—Isle of May, 19th March. *Autumn* 1883.—Isle of May, seen on 31st Oct., and several on 1st Nov., with other migrants.—*Reports on Migration of Birds*, 1879-85.

[1] See Yarrell's *British Birds*, 4th ed., vol. ii. pp. 274-277; also Seebohm's *British Birds*, vol. i. p. 544.

[2] Mr. Kelly records that they are occasionally found inter-breeding in Lauderdale.—*Hist. Ber. Nat. Club*, vol. vii. p. 304. Mr. George Bolam states that he has seen a mixed breed reared in the sea-cliffs to the north of Berwick, and that in 1880 he kept one of the young birds in confinement. It ultimately assumed a plumage half way between the two forms.—*Hist. Ber. Nat. Club*, vol. x. p. 391.

[3] In the spring of 1882 numbers of young chickens were taken away from the poultry yard at Nabdean by Carrion Crows, which were so cunning that they eluded most of the attempts made to shoot them. Mr. Hardy records (*Hist. Ber. Nat. Club*, vol. x. p. 559) that, in the same season, they carried off chickens from the Pease Mill.

Like the Raven, the Carrion Crow is omnivorous, but animal food of any kind is preferred by it, especially carrion.

> The toil more grateful as the task more low;
> So carrion is the quarry of the Crow.
>
> MALLET.

At the sea-side in winter it searches the shore for dead fish, and frequents the outer reefs at low tide to secure what food it can find there. During the severe winters of 1878-1881, when great numbers of Fieldfares, Redwings, and other small birds about the sea-coast perished from cold and hunger, their emaciated bodies were eagerly devoured by Carrion and Hooded Crows, which hovered round waiting for their prey.

The nest is usually placed in a Scotch fir, or in the fork of a tall ash or elm in a plantation. I have frequently seen it in the neighbourhood of Paxton in strips near the Tweed and the Whitadder. It is usually built of sticks and roots, and lined with moss, wool, and other soft materials. The eggs, which are from four to six in number, are very like those of the Rook, but larger; being generally bluish green, spotted and blotched with olive-brown. The Carrion Crow is a comparatively late breeder, and is found nesting in April and May. It is usually very wary when sitting on the nest, slipping quietly off long before a person can get within gunshot, unless when the eggs are nearly hatched. It may easily be distinguished from the Rook by the black feathers which cover the parts at the base of the beak, these parts in the adult Rook being bare of feathers.

This bird did not escape the notice of the superstitious in the county in olden times. Mr. Hardy writes that "It was once believed that when people played at cards on Sunday morning after a late Saturday night, the devil came

THE CARRION CROW.

down in the shape of a Black Crow, and sat on the table."[1]
Mr. Andrew Balsillie, Dykegatehead, informs me that he has heard it said that if a Hoodie be seen near a house in which some one is lying very ill, the sick person will not recover.

---

[1] Mr. Hardy's *MS. Notes.*

THE HOODED CROW.

GREY CROW, HEEDY CRAW, GREY-BACKED CROW, ROYSTON CROW, DUN CROW, BUNTING CROW, CROW.

Corvus cornix.

𝕿𝖍𝖊 𝕳𝖔𝖔𝖉𝖎𝖊, 𝕿𝖍𝖊 𝕳𝖚𝖉𝖉𝖎𝖊 𝕮𝖗𝖆𝖜.

The Huddit Crawis cryit varrock, varrock.
          SIR DAVID LINDSAY, *Complaynt of Scotland.*
The Rukis him rent, the Ravynis him druggit,
The Huditt Crawis his hair furth ruggit.
          W. DUNBAR.

As already mentioned in the account of the Carrion Crow, the Hooded Crow is now considered by ornithologists to be only a form of that bird. It is almost wholly migratory in Berwickshire, generally arriving from the north of Europe in October, and departing northwards in March and April;[1] but it has occasionally been known to remain during summer and breed in the county, when it has been sometimes seen with a Carrion Crow as a mate. Mr. H. H. Craw, Rawburn, near Longformacus, informs me that a pair of Hooded Crows remained on that farm in the summer of 1886, and nested and reared their young in a plantation called Allergrain Wood. It generally frequents the neighbourhood of

---
[1] The Hooded Crow is frequently seen on migration in spring and autumn at the lighthouses on the Isle of May and the Farne Islands.—*Reports on the Migration of Birds*, 1879-85.

the sea-shore, feeding on what the tide throws up, and on this account it is sometimes called the "Sea Craw." It is also seen by the side of the Tweed, where it searches for dead salmon and other garbage. When it comes inland, it frequents moorlands and open fields; Lauderdale, and the neighbourhood of Westruther and Spottiswoode, being some of its favourite resorts.

The back and under-parts of this species are grey, while its head, wings, and tail are black.

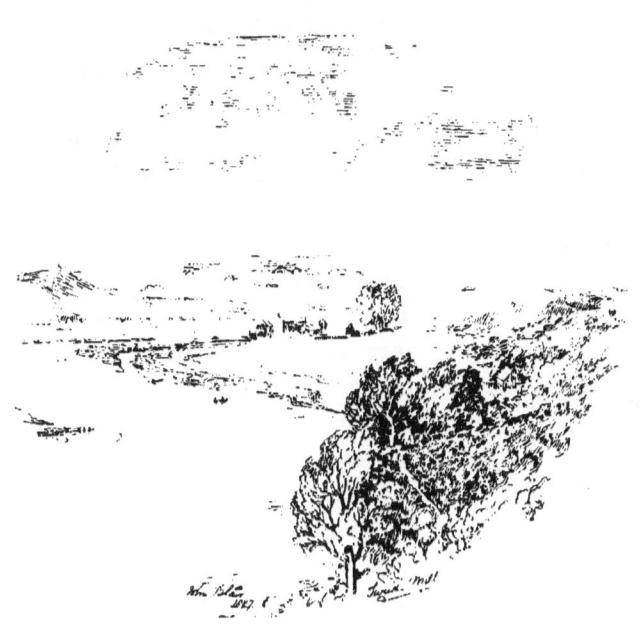

THE ROOK.

CROW, CRAW, CORN CRAW.

Corvus frugilegus.

The Craw, The Corn Craw.

Should I my steps turn to the rural seat,
Whose lofty elms and venerable oaks
Invite the Rook, who high amid the boughs,
In early Spring, his airy city builds,
And ceaseless caws amusive.
THOMSON, Spring.

Rooks are very numerous in Berwickshire, where they seem to have increased considerably within the last twenty or thirty years. The farmers complain of the damage which they do to newly sown corn and potato fields in spring, as well as to ripening grain in harvest. After the corn is cut, they attack the "stooks"; and, during severe weather in winter and spring, they commit depredations on stacks by making large holes in the thatch, to feed upon the ears of corn, which they carry off in great quantities, if they are allowed to do so. During the very stormy and snowy weather which was experienced in March 1888, they did much injury to ricks standing in fields at a distance from farm-steadings, and were sometimes seen feeding on them in thousands. When a "lying storm" of snow sets in, they frequent turnip fields, and pierce holes in the bulbs, which are rotted by water collecting in the holes and freezing. Wood Pigeons also get more easily at the inside of the bulbs when their skins are thus broken.

Rooks appear to have done so much damage to corn in the time of James I. of Scotland, that an Act was passed by the Scottish Parliament in 1424, for the purpose of lessening their numbers: "For thy that men considderis that Ruikes biggand in Kirk Zairdes, Orchardes, or Trees, dois greate skaith upon Cornes: It is ordained that they that sik Trees perteinis to, lette[1] them to big, and suffer on na wise that their birdes flie away. And quhair it be tainted that they big, and the Birdes be flowin, and the nest be funden in the Trees at Beltane,[2] the trees sal be foirfaulted to the King (bot gif they be redeemed fra him, throw them that they first perteined to), and hewin downe, and five schillings to the Kingis unlaw."[3]

This Act was followed by another of a like nature in 1427. Parliaments have at various times made similar laws for the destruction of many of our birds, including some which are now rare. In the time of Henry VIII. an Act was passed against Choughs, Crows, and Rooks, and the last-named birds were condemned by Elizabeth.[4]

> Some statis are plagued with snakis and frogs,
> And other kingdoms with mad dogs—
> Some are hurt with flocks of Crowes,
> Devouring corn and their lint bowes.[5]
>
> CLELAND.

Mr. George Buchan-Hepburn of Smeaton, in his *General View of the Agriculture of East-Lothian*, 1794, says that "Since the Earl of Haddington's woods at Tynningham have

---

[1] Hinder or prevent.

[2] Beltane, Beltein,—the name of a sort of festival observed on the 1st of May (O.S.) Hence used to denote the term of Whitsunday.—Jamieson's *Scot. Dict.* [This date would be about the 13th May at present.]

[3] *The Laws and Acts of Parliament of Scotland*, by Sir Thomas Murray of Glendook, Knight and Baronet: Edinburgh, 1681, p. 3.

[4] See Bishop Stillingfleet's *Works*, vol. ii. p. 561.

[5] Jamieson in his *Scot. Dict.* states that "lint bowes" are the globules which contain the seed of the flax.

grown into forest trees, the Crows have increased there to an astonishing degree, and they have really become a destructive nuisance. In the year 1775 a considerable number of the tenants in the neighbourhood of these woods associated together for the purpose of killing the Crows, and they assessed themselves at the rate of 5s. sterling per plough, and latterly at the rate of 2s. per ditto. Out of this fund they paid a bounty of a penny a head for old Crows; and in the beginning of the season they paid 2d. per dozen for the young ones; and as the season advanced, they increased the bounty to 3d., 4d., and 6d. per dozen. The numbers killed between 1779 and 1793 were 17,386 old Crows, and 59,269 young Crows, or in all 76,655, at an expense of £142, 14s. 7d. Mr. Robert Dudgeon, tenant in Tynningham, who paid the bounty and kept the accounts, furnished me with the above particulars; and he made the following observation, viz., that the expense of the killing of 76,655 Crows, amounted to a trifle short of 38s. per thousand,—whereas, if the damage done by a Crow in one year be estimated at 1d. only, 1000 Crows commit a waste, in that ratio, of nearly four guineas a year."[1]

Mr. Clay, Winfield, informs me that he has known Rooks pull up three or four acres of young turnips after they were singled; but this appears to be done to get at the grub at the root of the plant, in the same way as they sometimes dig up grass in pastures. It is believed by many naturalists that where the numbers are kept within reasonable bounds, the good which they do by destroying innumerable grubs and other noxious insects, outweighs the evil. If destructive insects, such as the caterpillar of the grass or antler moth (*Charæas graminis*), which lately appeared in vast num-

---

[1] *General View of the Agriculture of East-Lothian*, Edinburgh, 1794, pp. 141-143.

bers on some upland farms of the Borders, and destroyed acres of pasture, were not kept in subjection by Rooks, their ravages would soon become much more serious to the farmer than any damage done by these birds to his crops.[1] Mr. Hardy mentions that they are often very beneficial to oak trees in summer, by clearing their foliage of the caterpillar of the *Tetrix* and winter moth.[2]

This bird is omnivorous, but it appears to prefer worms and grubs, when these can be got, to any other food. Besides attacking farm crops when its favourite diet is scarce, it also destroys game by searching for and devouring the eggs[3] of Pheasants and Partridges, and likewise preying upon their young when newly hatched. It occasionally commits depredations amongst very small chickens in the poultry yard, an instance of this having occurred at Paxton in the spring of 1882, when many chickens were taken away by Rooks from the home farm of Nabdean. They sometimes pounced down on their prey within a short distance of the henwife, and were so wary that the gamekeeper, who kept watch with his gun, succeeded in shooting very few of the depredators.

Towards the end of summer, Rooks frequent the Lammermuirs, where they feed upon crawberries (*Empetrum nigrum*), and blaeberries (*Vaccinium myrtillus*).[4] They sometimes roost all night on the heather.

They seem to be very regular in some of their habits, for, at Paxton, where there is a large rookery by the side of

---

[1] For a full account of the ravages of the grass or antler moth on the Borders, see Mr. Hardy's paper on this insect in the *History of the Berwickshire Naturalists' Club*, vol. xi. pp. 195-205.

[2] Mr. Hardy's *MS. Notes*.

[3] About ten years ago a Rook was caught in a trap set at a Call Duck's nest at Paxton. It had eaten five of the eggs before the trap was put down, and when caught, the culprit was in the act of breakfasting on the remainder of the eggs in the nest.

[4] Mr. Hardy's *MS. Notes*.

the Tweed, and where they do not roost at night in winter, but come to the trees in the morning, and leave in the evening, I have observed that they come to the rookery about nine o'clock in the morning during that season, and towards three o'clock in the afternoon return in straggling flocks to the west, probably on their way to their winter roosting-trees at Millburn, Marchmont, or Mellerstain.

> November chill blaws loud wi' angry sough,
> The short'ning winter day is near a close;
> The miry beasts returning frae the pleugh:
> The black'ning trains o' Craws to their repose.
> BURNS.

Were it not for the annual "Crow-shootings" which take place at the various rookeries in the county, about the second week in May, when many thousands of the young birds are killed, Rooks would soon increase to such an extent that their natural food would fail, and they would commit wholesale destruction amongst the crops of the farmer. In the rookery at Paxton, which is only of moderate extent, no fewer than 2400 young birds have been killed in a season. A little calculation will show the vast increase which would take place in the number of rooks in the county, if even the young birds at Paxton Rookery only were allowed to "fly" annually. It is therefore not surprising in early times, when corn crops were scanty in Scotland, and guns were not readily available for the destruction of the young birds, that, as already mentioned, an Act of Parliament was passed in the fifteenth century, to force proprietors of rookeries to harry the nests before the young could fly.

The following list of Rookeries in Berwickshire has been prepared from answers to circulars sent by me to every parish in the county, in February 1887:—

Parish.	Situation of Rookery.	Kinds and age of trees.	Size of Rookery.	Name and Address of Reporter.	Remarks.
Abbey St. Bathans.				James Shiel, gamekeeper, Abbey St. Bathans.	There are no rookeries in this parish. — James Shiel, Abbey St. Bathans, Grantshouse.
Ayton.	Ayton Castle, on the banks of the Eye.	Ash, elm, and beech.	Over 500 nests.	Peter Scott, farmer, Whiterig, Ayton.	Rooks seem to have increased in the parish within the last twenty years. — Peter Scott, Whiterig, Ayton.
	Peelwalls.	Principally ash of large size.	About 200 nests.	Do.	Rooks do not roost at night during the winter months at Ayton Castle or Peelwalls. They roost there at night only during the breeding season. — Robert Steele, gamekeeper at Ayton Castle, April 22, 1888.
	Whiterig.	Ash, over 100 years old.	About 100 nests.	Do.	
	Ayton Cocklaw Farm, near Lamberton Moor.		About 200 nests.	Robert Campbell-Renton, younger of Mordington, Berwick-on-Tweed.	
	On a field of Ayton Cocklaw Farm, near Ayton Station.	A few nests.		Do.	
Bunele.	Near Blanerne House, and in the policy grounds there.	Very old elm, beech, oak, plane, and Scotch fir.	About 500 nests.	John Blackadder, East Blanerne, Chirnside.	The Rooks roost at night at Blanerne in the end of August and September, in calm weather. During stormy weather they roost at the rookery near Chapel Farm, in Duns parish. — [The Peely Braes.] In former times there were rookeries at the ruins of Bunele Castle and at Billie Castle.

Parish.	Situation of Rookery.	Kinds and age of trees.	Size of Rookery.	Name and Address of Reporter.	Remarks.
Buncle—continued.)					Rooks have greatly increased in this parish since the imposition of the gun tax.—John Blackadder, East Blanerne. Rooks do not roost at Blanerne at night during the winter, but pass on to Barrowmill Woods on Oxendean Estate.—[The Peely Braes.] They begin to roost at Blanerne at night in March, and leave off doing so in the autumn.—R. Bruce, gamekeeper, Blanerne.
Channelkirk.				Robert Romanes, Harryburn, Lauder.	There are no rookeries in this parish.
Chirnside.	Whitehall.	Oak, elm, and ash, at least 150 years old.	500 nests and upwards.	Charles Stuart, M.D., Hillside Cottage, Chirnside.	Rooks roost at Whitehall at night in the winter months. They have increased there since the gun tax was imposed.
	Ninewells.	Scotch fir, elm, oak, about 200 years old.	From 200 to 500 nests.	Do.	Rooks roost at Ninewells at night in the winter months.
	Maines.	Ash, elm, and various other kinds of trees.	From 50 to 200 nests.	Do.	Charles Stuart, M. D., Hillside Cottage, Chirnside.
	Near Chirnside Mause.		A few nests.	Do.	

Parish.	Situation of Rookery.	Kinds and age of trees.	Size of Rookery.	Name and Address of Reporter.	Remarks.
Cockburnspath.	On Dean Burn (part of Dunglass Rookery).	Firs and deciduous trees of a great height.	About 400 or 500 nests. — [John Bolton, gamekeeper, Dunglass.]	James Hardy, Oldcambus, Cockburnspath.	The rookery at Dunglass is very large, consisting of about 800 or 900 nests. About half of the nests are on the Berwickshire side of the Dean Burn. Rooks are increasing much in number in this parish. — James Hardy, Oldcambus. Rooks roost in thousands in the Dunglass Dean rookery all through the winter months. A far greater number roost there at that season, than breed there in summer. They appear to come in to roost from the neighbouring rookeries in the winter afternoons.— J. Bolton, gamekeeper, Dunglass.
Coldingham.	Houndwood.	Elm, ash, and beech.	From 250 to 300 nests.	Dr. James M'Dougall, Coldingham.	The Rooks do not appear to roost at any rookery in this parish, at night, during the winter months. They leave for Whitehall in November and December, and return in February. They sometimes visit the rookeries during the day in winter time. About 30 years ago there was a rookery at Coldingham Dean, consisting of 300 or 400 nests, but the trees were cut down.
	Templehall.	Elm, ash, and beech.	Small.	Do.	
	Berrybank.		About 100 nests here. This rookery was begun only 17 years ago.	Do.	

Parish.	Situation of Rookery.	Kinds and age of trees.	Size of Rookery.	Name and Address of Reporter.	Remarks.
Coldingham, *continued.*	Bogangreen.	Elm, ash, and beech.	About 70 nests here; this rookery has been in existence many years.	Dr. James M'Dougall, Coldingham.	Miss Coulson of Houndwood, states that the rookery now at Houndwood was formed about fifty or sixty years ago, when Rooks were banished from Netherbyres; and between 400 and 500 young Rooks are shot there every season. The rookeries suffered from the great gale in October 1881. Since then the Rooks have been on the increase in three of the four rookeries in the parish.—Dr. James M'Dougall, Coldingham.
	Greenburn.	Ash, about 50 years old.	8 to 10 nests.	John Blackadder, East Blanerne, Chirnside.	Greenburn rookery was formed four years ago.—John Blackadder, East Blanerne.
Coldstream.	Milne Graden.	Beech, ash, elm, oak, and Scotch fir, very old.	796 nests.	Jas. Graham, gardener, Milne Graden, Coldstream.	Rooks roost at Milne Graden rookery throughout the year, including the winter months. —James Graham, Milne Graden. Rooks have increased in the parish within the last 20 years.— J. White, The Hirsel.
	Castlelaw.	Ash, oak, beech, about 150 years old.	About 500 nests.	J. White, gamekeeper, The Hirsel, Coldstream.	
	The Hirsel.	Do.	About 100 nests.	Do.	
	Hatchednize.	Ash, oak, about 100 years old.	About 100 nests.	Do.	
	The Lees.		Small.	Do.	

Parish.	Situation of Rookery.	Kinds and age of trees.	Size of Rookery.	Name and Address of Reporter.	Remarks.
Cranshaws.				John S. Bertram, farmer, Cranshaws, Duns.	There are no rookeries in this parish.
Duns.	The Peelies on the south bank of the Whitadder.	Ash, elm, and other hard woods, over 100 years old.	Upwards of 1000 nests.	John Ferguson, Duns.	Rooks roost at night time at the Peelies during the winter months in great numbers. The average number cannot easily be estimated, but it cannot be less than 4000. Rooks have apparently slightly increased in the parish within the last twenty years.—John Ferguson, Duns.
	Cumledge.		About 100 nests.	David Richardson, gamekeeper at Broomhouse, Duns.	
Earlston.	At Mellerstain.	Beech, ash, and Scotch fir, probably 150 or 200 years old.	The rookery is about 1 mile long, and there are in it from 1200 to 1500 nests.	Jas. MacPherson, gamekeeper, Mellerstain.	The Rooks roost at Mellerstain all the year round. They are more numerous in the winter than in summer. They seem to gather in from other districts in winter to roost. They do not roost in winter at Haughhead. No decrease has been observed in the number of Rooks in the parish for the last twenty years. — James MacPherson, Mellerstain, Kelso.
	Near Haughhead.	Ash and fir.	Between 70 and 100 nests.	Do.	
Eccles.	Belchester.	Mixed hardwood, about 150 years old.	500 nests and upwards.	Charles Waddell, Birgham Haugh, Coldstream.	There used to be a rookery at Fernierig, of 20 to 50 nests on elm and plane trees. Most of

VOL. I.   P

Parish.	Situation of Rookery.	Kinds and age of trees.	Size of Rookery.	Name and Address of Reporter.	Remarks.
Eccles— continued.	Bartlehill.	Mostly elm.	200 nests and upwards.	Charles Waddell, Birgham Haugh, Coldstream.	the trees were blown down by the great gale of October 1881, and since then the Rooks have abandoned the place. Rooks have increased in the district, if not in the parish, within the last twenty years.—Charles Waddell, Birgham Haugh. The Rooks roost in Belchester, Kames, and Stainrig rookeries, at night, in the winter months.—Robt. Gillon, gamekeeper, Antonshill, Coldstream.
	Captain's Plantation, near Bartlehill.	Elm, spruce, Scotch fir, and birch, about 80 years old.	200 nests and upwards.	Do.	
	Eccles.	Old elm and beech.	500 nests and upwards.	Do.	
	Eccles, Newtown.	Ash and elm, over 100 years old.	50 to 200 nests.	Do.	
	Kames.	Various kinds, 60 or 70 years old.	500 nests and upwards.	Do.	
	Mersington.	Mostly elm.	500 nests and upwards.	Do.	
	Purveshall.	Various kinds, very old.	500 nests and upwards.	Do.	
	Stainrig.	Hard wood, very old.	500 nests and upwards.	Do.	
Edrom.	Blackadder.	Principally beech and fir, over 70 years old.	About 1000 nests.	Rev. MacDuff Simpson, The Manse, Edrom.	There was at one time a rookery at Broomhouse, but the trees were cut down. There was also a small one at Edrom, but the Rooks were all banished from it by continual shooting. Rooks have increased in the parish within the last 20 years.—Rev. M. Simpson, Manse, Edrom.

THE ROOK.

PARISH.	Situation of Rookery.	Kinds and age of trees.	Size of Rookery.	Name and Address of Reporter.	REMARKS.
Edrom—continued.	Nisbet House.	Beech, fir, and elm, from 50 to 100 years old, and within a circle of half a mile of Nisbet House.	About 500 nests.	Rev. MacDuff Simpson, The Manse, Edrom.	Rooks do not roost at Blackadder at night in winter. They give up roosting in October and commence to do so again in spring when they rebuild their nests.—J. Hogg, gamekeeper, Blackadder, Chirnside.
	Kimmerghame.	Beech, about 80 years old.	About 400 nests.	Do.	Rooks do not roost at Kimmerghame at night in winter. They come to the rookery in February, and remain until the end of September or later.—John Thomson, gamekeeper at Kimmerghame, Duns.
	Kelloe.	Beech and fir, about 75 years old.	About 200 nests.	Do.	Rooks roost at Kelloe all the winter months.—Hugh Frazer, gamekeeper at Kelloe, Edrom.
	Near Allanton Free Church.		Small—it is only 4 or 5 years old.	Do.	
Eyemouth.				J. Donaldson, Eyemouth.	There are no rookeries in this parish. In 1884 a pair of Rooks built a nest on a tree near Gunsgreen House. When it was about finished they spent a day in what appeared to be quiet observation, and then all at once removed the nest to the vicinity of Ayton Castle. Probably a study of the faces of the boys of Eyemouth led to this action.—J. Donaldson, Eyemouth.

Parish.	Situation of Rookery.	Kinds and age of trees.	Size of Rookery.	Name and Address of Reporter.	Remarks.
Fogo.	Chesters.	Ash, elm, & spruce, from 50 to 60 years old.	About 60 to 100 nests in April 1888.	George Falconer, gardener, Caldra House, Duns. W. Thompson, gamekeeper at Charterhall, Duns.	Rooks have increased in the parish within the last twenty years. There used to be a small rookery at Bogend, but it is now abandoned. — G. Falconer, Caldra House.
Foulden.	Foulden House.	Plane trees and beech, elm, and ash, supposed to be from 250 to 300 years old.	Upwards of 500 nests.	H. Hewat-Craw, farmer, Foulden West Mains, Foulden, near Berwick-on-Tweed.	Rooks do not roost at Foulden at night during the winter months. They give up roosting at night in the end of August or September, and commence to roost at night time in the beginning of March. They are believed to roost mostly at Whitehall during the winter months. Rooks have considerably diminished in the parish during the last twenty years.— H. Hewat-Craw, Foulden West Mains.
	Nunlands.	Plane, supposed to be from 250 to 300 years old.	Under 20 nests.		
Gordon.				Robert Renton, Greenlaw.	There are apparently no rookeries in this parish. There used to be one in a wood called Hunny Crooks, to the south of Rumbleton-Law, but the trees were cut down when the railway was made.— Robert Renton, Greenlaw.

THE ROOK.

Parish.	Situation of Rookery.	Kinds and age of trees.	Size of Rookery.	Name and Address of Reporter.	Remarks.
Greenlaw.				Peter Loney, Marchmont, Duns.	There are no rookeries in this parish. There was a rookery at Bedshiel—about 30 nests—twenty years ago, but the trees have all been blown down. — Peter Loney, Marchmont.
Hume.	Humehall.	Elm and ash, from 60 to 80 feet high.	About 500 nests.	Peter Loney, Marchmont, Duns.	Rooks seldom roost at Humehall at night during the winter months. Rooks do not roost at Todrig at night in winter. The numbers of rooks in the parish have continued about the same for the last twenty years.—Peter Loney, Marchmont.
	Todrig.	Ash.	Small.	Do.	
	Fallsidehill.		Small.	J. M'Pherson, gamekeeper, Mellerstain, Kelso.	
Hutton.	In the policy ground of Paxton.	Ash, elm, & beech, spruce & Scotch fir, about 100 years old.	742 nests in 1887.	William Preston, gamekeeper, Paxton House, Berwick-on-Tweed.	Rooks do not roost at night at the Paxton Rookery during the winter months. At that season they only visit the rookery during the day, coming from the west in the morning about nine o'clock, and returning westwards about three o'clock in the afternoon.
	Hutton Hall.		About 50 nests.	Robert Nesbit, farmer, Hutton Hall Barns, Hutton, Berwick-on-Tweed.	

Parish.	Situation of Rookery.	Kinds and age of trees.	Size of Rookery.	Name and Address of Reporter.	Remarks.
Hutton— continued.	Near Bluestane Ford.		About 14 nests.	Robert Nesbit, farmer, Hutton Hall Barns, Hutton, Berwick-on-Tweed.	The Rooks begin to roost at night in March. Rooks have increased in numbers in the parish within the last twenty years.— William Preston, gamekeeper, Paxton House.
Ladykirk.	In the policy grounds of Ladykirk House.	Oak, ash, elm, beech, and plane.	From 500 to 1000 nests.	John Miller, Ladykirk West Lodge, Norham.	There are two rookeries in the policy grounds of Ladykirk House — one called the "Craw Dean," and the other near the West Lodge. The Rooks roost at night during the winter months at Ladykirk. Rooks have increased in the parish during the last twenty years. — John Miller, Ladykirk West Lodge.
Langton.	Near Langton House.	Beech, ash, and plane, over 100 years old.	About 400 nests.	John Ferguson, Duns.	The Rooks roost at Langton at night during the winter months. The number of Rooks in the parish appears to be stationary.— John Ferguson, Duns.
Lauder.	Eagerhope Wood.	Scotch fir, ash, and plane, on the brae face and in glens; old.	500 nests and upwards.	Robert Romanes of Harryburn, Lauder.	Rooks have not increased much in the parish within the last twenty years.

THE ROOK.

Parish.	Situation of Rookery.	Kinds and age of trees.	Size of Rookery.	Name and Address of Reporter.	Remarks.
Lauder— *continued.*	Allanbank.	Ash, elm, plane, and fir on level ground, 53 years old.	From 200 to 500 nests.	Robert Romanes, Harryburn, Lauder.	The rookery at St. Leonards is decreasing, while that at Thirlstane is increasing rapidly. The rookery at Thirlstane was destroyed by the gamekeepers exterminating the Rooks. They have returned after a lapse of fifty years. It is thought that the Rooks do not roost at night time during the winter months in any rookery in the parish.— Robert Romanes, Harryburn.
	Chapel.	Scotch fir, ash, and elm, on brae face, and in long shelter strip.	From 200 to 500 nests.	Do.	
	St. Leonards.	Ash, elm, and plane, on brae face.	50 to 200 nests.	Do.	
	At Thirlstane Castle policy, behind the lower part of Lauder.	Ash, plane, and elm.	50 to 200 nests.	Do.	
	The strip between Hogg's Blainslie and St. Leonards.	Scotch fir.	20 to 50 nests.	Do.	
	Cauldshield. Little plantation in middle of field.	Scotch fir.	Under 20 nests.	Do.	
Legerwood.	Corsbie Tower.	Beech.	About 200 nests.	Walter Lockie, teacher, Gateside, Spottiswoode, Westruther.	Rooks do not roost in these rookeries at night during the winter months. There used to be a rookery at Birkhillside about 40 years ago, but trees were taken down. The numbers of Rooks in the parish have remained stationary for the last 20 years.—W. Lockie, Gateside.
	Corsbie Stell.	Fir.	20 to 50 nests.		

Parish.	Situation of Rookery.	Kinds and age of trees.	Size of Rookery.	Name and Address of Reporter.	Remarks.
Longformacus.	In the policy grounds of Longformacus House.	Ash, beech, elm, plane.	500 nests and upwards.	Rev. George Cook, The Manse, Longformacus.	There are four rookeries at Longformacus House, within a radius of a mile. The Rooks roost at the Longformacus Rookery the whole year round, including the winter months. They are more numerous in severe weather. Rooks have increased in the parish during the last twenty years. — Rev. George Cook, The Manse, Longformacus, Duns.
Mordington.	Near Mordington House.	Plane, ash, elm, from 100 to 300 years old.	There are 3 rookeries of about 300 nests each.	Robert Campbell-Renton, yr. of Mordington, Mordington House, near Berwick-on-Tweed.	Rooks have largely increased in numbers in the parish within the last twenty years. One rookery, now containing 300 nests, has been in existence for only five years. About a thousand young Rooks were shot in May 1887.— Robt. Campbell-Renton, Mordington House. The Rooks do not roost at Mordington at night during the winter months. They begin to roost in March, and give up doing so when the leaves begin to fall in the autumn.—James Purves, gamekeeper, Mordington.

Parish.	Situation of Rookery.	Kinds and age of trees.	Size of Rookery.	Name and Address of Reporter.	Remarks.
Mertoun.	Bemersyde. In the policy ground of Mertoun.	Various kinds of trees, including elm, Scotch fir, and ash.	About 200 nests. Over 500 nests.	John Thomson, Maxton, St. Boswells.	Rooks do not roost at Mertoun at night during the winter months. They have increased greatly in the parish during the last twenty years.—John Thomson, Maxton.
Nenthorn.				John Black, farmer, Girrick, Kelso.	There are no rookeries in the parish of Nenthorn. About ten years ago a few nests were built on large lime trees at Nenthorn House, but they were blown down and never rebuilt. The Rooks have greatly increased in the district within the last twenty years.—John Black, Girrick.
Oldhamstocks. (Part in Berwickshire.)				James Hardy, Oldcambus, Cockburnspath.	There are no rookeries in this part of Oldhamstocks parish.—James Hardy, Oldcambus.
Polwarth.	Marchmont.	On Spanish chestnut, plane, ash, elm, beech, and yew.	500 nests and upwards.	Peter Loney, Marchmont, Duns.	The Rooks roost at night all through the winter at Marchmont.

Parish.	Situation of Rookery.	Kinds and age of trees.	Size of Rookery.	Name and Address of Reporter.	Remarks.
Polwarth— continued.		The trees are large, many of them being 100 feet high.			The Rooks are more numerous then than during the breeding season. The numbers of Rooks have kept much about the same during the last 20 years.—P. Loney, Marchmont.
Swinton.	Near west end of Swinton village.	Old oak and beech.	About 40 nests in 1886.	Geo. Tweedie, teacher, Swinton, Duns.	The Rooks do not remain to roost at night after the breeding season is over. The rookery was begun in 1880 or in 1881, with a few nests only. Rooks appear to have increased in the parish within the last twenty years. — George Tweedie, Swinton.
Westruther.	Spottiswoode, from Gateside to Spottiswoode West Lodge, and Dumbhouse wood, adjoining.	Principally Scotch fir and spruce, over 100 years old.	Between 500 and 1000 nests.	Walter Lockie, teacher, Gateside, Spottiswoode, Westruther.	The Rooks roost at Spottiswoode the whole year round, except when a severe snow-storm occurs in winter, when they leave for other quarters. The numbers of Rooks in the parish are said to have been stationary for the last twenty years. — Walter Lockie, Gateside.
Whitsome.				Charles Stuart, M.D., Hillside Cottage, Chirnside.	There are no rookeries in the parish of Whitsome. — Charles Stuart, M.D., Chirnside.

The Rooks generally begin to repair their old nests or build new ones about the first week of March, and during the progress of the work great pilfering of the materials and fighting often takes place amongst the builders, the uproar in the rookery being very great all the while. The nests, which are constructed of sticks and twigs, lined with fibrous roots, straw, grass, and other soft materials, are generally placed towards the top of the trees in the rookery, as many as ten or fifteen being often seen on one tree. The eggs, which are from four to six in number, are bluish green, blotched, streaked, and spotted or freckled with olive-brown. The plumage seldom varies, but occasionally an albino occurs. A white-coloured specimen was got at Spottiswoode in 1883.[1]

The Rev. Alexander Waugh, D.D., London, was born on the 16th of August 1754, at East Gordon, in the Merse, and the following is one of his anecdotes connected with it: "I remember when I returned home at the vacation of Earlstoun school, I frequently went out to the Muir to have some talk with my father's shepherd, a douce, talkative, and wise man in his way; and he told me, a wondering boy, a great many things I had never read in my school-books. For instance, about the Craws—(there were plenty of Craws about Gordon Muir, and I often wondered what they got to feed on), that they aye lay the first stick of their nests on Candlemas day, and that some of them that big their nests on rocks and cliffs have siccan skill of the wind that if it is to blow mainly from the east in the following spring, they are sure to build their nests on what will be the bieldy side, and many a ane that notices it can tell frae what airt the wind will blaw. After expressing my admiring belief

---

[1] See *Scotsman*, 3rd March 1887.

of this, I thought, as I had begun Latin, and was therefore a clever chield, that I wadna let the herd run away wi' a' the learning. It was at the time when the alteration in the style had not ceased to cause grief and displeasure to many of the good old people in Scotland, and I knew the herd was a zealous opponent of the change, so I slily asked him, 'Do the Craws count Candlemas by the new or the auld style?' He replied, with great indignation—'D' ye think the Craws care for your Acts of Parliament?"[1]

The "building o' the Craws" is one of those rural events which never fail to attract attention, and it is a popular saying in Berwickshire that

> On the first of March,
> The Craws begin to search

for sticks and other materials for their nests. Boys in the county are sometimes puzzled by being asked, "How many sticks gang to the building o' a Craw's nest?" and seldom give the required answer, which is, "Nane; they're a' carried." The following fragment of an old rhyme is also occasionally heard—

> Some gaed east, and some gaed west,
> And some gaed to the Craw's nest.

When children see Crows hastening away in a body, they say, "the schule is skailin'," and if they are noticed sitting thickly on trees in the winter time, it is called "a Craw's preachin'." When they are seen wheeling and hovering round one spot in numbers, old people say it is "a Craw's waddin"—

> A weddin' o' Craws, a weddin' o' Craws,
> A piper, and a fiddler, and three Jackdaws.

If they are observed flying high in the air in a flock, and tumbling and diving down, it is considered a sign of wind

---

[1] *Memoir of the Rev. Alexander Waugh, D.D.*, London, 1830, pp. 390-391.

or rain.¹ If Crows come near dwelling-houses in considerable numbers during winter, in search of food, especially about the doors, it is alleged that a severe gale is brewing, and will speedily arrive.² In winter, when they are observed flocking and feeding eagerly in the fields, if there be no snow on the ground, a snow-storm is thought to be indicated—

> Now cauld blaws the wind and the north's angry form
> Spreads dark owre the welkin around ;
> The bizzie Craws tell o' the fierce coming storm,
> Where thousands feed black'ning the ground.³

Should snow be lying on the ground at the time, fresh weather is anticipated. When they sit silent in rows on palings and dykes, rain is thought to be foretold.⁴ The apparent susceptibility of the Rook to changes of the weather seems to have attracted the notice of Virgil, and is thus alluded to in his *Georgics* :—

> Tum liquidas corvi presso ter gutture voces
> Aut quater ingeminant ; et sæpe cubilibus altis,
> Nescio quâ præter solitum dulcedine læti,
> nter se foliis strepitant ; juvat, imbribus actis,
> Progeniem parvam dulcesque revisere nidos.

---

[1] Mr. Lockie writes on 29th Jan. 1887, with regard to the neighbourhood of Spottiswoode :—"If Crows are observed high overhead in a flock, and begin suddenly to dive and sweep towards the earth, it is said that a high wind will soon spring up. This is a common remark made during the harvest season when crops are in stook, as farmers look eagerly for every sign of a drying wind during that fickle period." Mr. James Smith, shepherd, Byreclough, told me in July 1886, that in that district, when Crows are seen flying high in the air and diving down, it is believed to presage wind and rain. The following is an extract from the Meteorological Registers of Captain Bell, of the Berwickshire Militia, who lived at Linthill, near Eyemouth, in 1802 :—"Linthill, 7th Oct. 1802.—The Crows flying and 'swiping' down in different directions, as they do before wind." On the 8th he notes :—"It blows a good deal."

[2] *Spottiswoode District.*—W. Lockie.

[3] *Poems of Andrew Scott :* Bowden, 1801.

[4] Miss Georgiana Milne-Home has informed me that, on some occasions, when the Rooks return to the rookery at Milne Graden in the autumn and winter afternoons, instead of alighting upon the trees in the usual way, they all sit down in immense flocks in the adjoining fields, and wait there, apparently doing nothing, until it is almost dark. I have heard it said that they have been observed to do this on the approach of stormy weather.

When a Crow, by some mischance, gets its plumage wet, the bedraggled appearance which it presents has given rise to the expression "drookit like a Craw," which is often heard in the county in rainy weather with regard to a person who has been exposed to a heavy downfall, and whose clothes have got soaked :—

> Wet to the skin amang the stooks,
> We gang a' day wi' waesome looks,
> Like drookit Craws or heartless Rooks,
>   The sheaves to set up,
> And lang for night and cozy nooks,
>   And hearths weel het up.[1]

About the nesting season some Rooks get so hoarse that they cannot utter their usual "craw," but give vent to an abnormal note instead; and in some cases no sound is heard at all, although the bird is seen to open its bill, and attempts to "craw" in the ordinary way. This peculiarity has apparently been the origin of the popular remark often heard with reference to a person who is very hoarse, that he is "roopit like a Craw," or "as hoarse as a Craw." "As black's a Craw" is a common expression in the county; and self-admiration is hinted at by the following proverb, "Ay, but ye're a bonnie pair, as the Craw said to its ain twa feet."[2] The expression "as straight as the Crow flies" is sometimes heard, and when the distance across country between two places is spoken of, it is said to be "as the Crow flies."

The following places in the county have seemingly derived their names from having been much frequented by Crows :—The Crow Wood, near St. Leonards, parish of Lauder; Crawlee, in Greenlaw parish; North and South

---

[1] "Lines on the late unfavourable weather," by George Henderson, Surgeon, Chirnside, 12th September 1857.

[2] This proverb was repeated to me by Miss Georgina Milne-Home, Milne Graden, 31st January 1887.

Crawlea Plantations, near Foulden Hill, Foulden parish; Crawlee Burn, Gordon parish; Crawlaw, a hamlet about a mile north from Wedderlie, in Westruther parish; South Crawhill, Easter Crowbutts, and Wester Crowbutts, Chirnside parish; Crowshiel, and Crow Green, on the Tweed, near Lennel, Coldstream parish; Craw Cleuch, on Kilpallet Burn, and Little Craw Cleuch, on a tributary of Kilpallet Burn, on the march with East-Lothian; also Craw Cleugh, on a tributary of Watch Water, about half-a-mile west from Twinlawford, in Cranshaws parish. Craw Burn is an old name of Howpark Burn, a tributary of the Eye, which is still applied to the under-part of it; Craw's Entry, leading from Bent's Corner plantation to Marchmont policy-ground, in Polwarth parish. East and West Crow Butts are likewise the names of two fields on the farm of Blackpotts, in the parish of Coldingham; and the "Rooks" and the "Little Rooks" are the names of two rocks off the coast at Dowlaw in the same parish.

Craw-taes (*Lotus corniculatus*), Craw-pease (*Lathyrus pratensis*), Craw-berry and Craw-crooks (*Empetrum nigrum*), are popular names of wild plants in Berwickshire.[1]

---

[1] Dr. Johnston, *Natural History of the Eastern Borders*, vol. i., Botany. pp. 55, 56, 175.

THE RAVEN.

CROW, CORBY OR CORBIE.

Corvus corax.

The Corbie.

> O_wre the moor, near yonder kirk,[1]
> We'll set the fagots in a lowe,
> And wrap the hags in tar and towe;
> And there we winna let them shirk,
> But scouther them wi' broom and birk,
> Bleezin' on the Witches' Knowe,[2]
> Bleezin' round ilk hoary pow,
> How the hags will girn and gape;
> Satan, in a Corbie's shape,
> Will come and take his pets away—
> Sic a bleeze we'll hae that day.
>
> POPULAR RHYMES OF BERWICKSHIRE.

IN former ages the Raven was considered to be a bird of ill-omen, and in Berwickshire it was of such evil repute, that it was even supposed that the devil himself sometimes appeared in the shape of a Corbie. It was also associated in the popular imagination with death, and dead bodies; and

---

[1] The old "kerke" of Lamberton, in the parish of Mordington. Here the Princess Margaret, daughter of Henry VII. of England, was betrothed to James IV. of Scotland, on 1st August 1502. A full account of the proceedings at the "kerke" on this occasion is given in Carr's *History of Coldingham Priory*, pp. 39-42. Only a small portion of the ancient walls is now standing.

[2] A small round hill about half-a-mile westward from the ruins of Lamberton Kirk. Here, tradition says, two witches were burned about the beginning of last century. They are supposed to have been the last who suffered for the crime of witchcraft previous to the enactment of George II. in 1722.—Carr's *History of Coldingham Priory*, p. 147.

this added to the hatred which was entertained towards the bird. In the troublous times of the Regent Albany, when the Chevalier de la Beauté, who had been appointed Warden of the Eastern Marches in the place of the Earl of Home, was fleeing from the Homes of Wedderburn, with whom he had quarrelled near Langton Tower, on the 20th of September 1517, he is said to have been warned by a "weird auld man" not to cross the streamlet which flows between Langton and Duns;[1] hence the popular rhyme—

> If ye pass owre the Cornysyke,
> The Corbies will get your bones to pyke.[2]

The following allusion to this bird is likewise made in a Berwickshire rhyme, referring to the atrocious murder of "Lady" Billie by her butler, Norman Ross, at Linthill House, near Eyemouth, on the 12th of August 1751—

> Norman Ross wi' pykit pow,
> Three Corbies at his e'en;
> Girnin in the gallows tow,—
> Sic a sight was never seen.[3]

In the olden times, when reiving was a favourite pursuit of the Borderers, the Raven seems to have occasionally come in for his share of the spoil.

> The Corbies in the Corbie Heugh [4]
> Are crouping like to dee;
> But our laird will gie them meat eneugh,
> And that you'll soon see,
> When Houldie and his reivers rude
> Hing on the gallows tree.

The late Dr. Henderson, of Chirnside, mentions in his

---

[1] For a full account of the quarrel between Home of Wedderburn and De la Beauté, see Godscroft's *History of the Homes of Wedderburn*.

[2] *Popular Rhymes of Berwickshire*, p. 20.   [3] *Ibid.* p. 128.

[4] There is, or was, a locality near Ayton called the Corbie Heugh, because numbers of Corbies were wont to breed there in former times.—Dr. Henderson's *Pop. Rhymes of Ber.*, p. 23.

Popular Rhymes, Sayings, and Proverbs of the County of Berwick, a remarkable superstition connected with a small stone said to have been taken out of a Corbie's nest at the Corbie Heugh near Ayton, to which locality the above rhyme refers. He says :—" Our great-grandfather lived in Ayton about 1730, and he got into his possession an article of glamourie which he took out of a Corbie's nest in the Corbie Heugh, and which is said to have wrought many miraculous cures both on man and beast." Amongst some manuscript notes of the late Dr. Henderson, which have been kindly given to me for reference by his son, Mr. Robert Henderson, Chirnside, I find the following wonderful account of this stone :—" It was in great repute for healing tumours, swellings, and sores, and also for curing distempers in cattle, horses, and sheep, and the Laird of Manderston, near Duns, once sent for it to cure his cattle of some desperate disease which raged amongst them, and which had carried off some of his best animals ; but after the Corbie Stone was laid into the pond where the cattle drank, not one of them died, and they all got better in a day or two."

There are several places in the county besides that above mentioned which have apparently taken their names from having been frequented by Ravens in former times, when the bird was comparatively common. Of these may be mentioned Corbie Hall, in the parish of Fogo ; Corbie Heugh Quarry, near the meeting of the Blackadder and its tributary the Fangrist, in Greenlaw parish ; Corbie Hill, Raven's Brae, and Raven's Heugh,[1] in the parish of Coldingham ; Raven's Knowe, on the Whitadder, near Edrington Mill ; and Raven's Craig, a precipice on the same river in the

---

[1] Part of Weston Thirl Brae, called Raven's Heugh, because the Raven used to build there. It is also known as Thomas Bookless' Heugh, from his having been killed there by a fall from the cliffs while harrying a Raven's nest.—J. Hardy's *MS. Notes*.

neighbourhood of Cranshaws. A popular rhyme referring to a superstition connected with the last named locality runs—

> It's no weel mow'd! It's no weel mow'd!
> Then it's ne'er be mow'd by me again,
> I'll scatter it owre the Raven's Stane,
> And they'll hae some wark e'er its mowed again![1]

A "Corbie messenger" is one who either does not return at all, or returns too late.

> Thou corby messinger, quoth he, with sorrow now sings.
> — HOLLAND'S *Houlat*, about 1453.

> He send furth Corby Messingeir
> Into the air for to espy
> Gif he saw ony mountains dry,
> Sum sayis the Rauin did furth remane
> And come not to the ark agane.
> — LINDSAY'S *Warkis*, 1592.

A Scottish proverb, referring to those of the same profession and the like, says: "Ae Corbie will no pyke out anither's een;" and in Berwickshire, Mr. Hardy says, "Fell on me like a Raven" used to be a common expression, when allusion was made to any savage attack.

Although this species does not now breed on the seacoast of Berwickshire, and is but seldom seen in the county, it appears to have been comparatively common long ago; its chief resort being the precipitous cliffs in the neighbourhood of Fast Castle and St. Abb's Head, where in some crevice of the rock, at a vast height above the boiling surge below, it placed its nest and reared its young.

---

[1] In old time Cranshaws was the habitation of an industrious *Brownie*, in so much that the barn-man's office became a perfect sinecure. This Brownie both inned the corn and thrashed it, and that for several successive seasons. It at length happened, one harvest, that after he had brought the whole victual into the barn, some one remarked that he had not *mowed* it very well—that is, not piled it up neatly at the end of the barn; whereat the spirit took such offence that he threw the whole of it next night over the *Raven* Craig, a precipice two miles off, and the people of the farm had almost the trouble of a second harvest in gathering it up again.—*Robert Chambers*.

The Rev. Andrew Baird, writing in 1834, mentions that the Raven then built in the most inaccessible cliffs on the sea-coast of the parish of Cockburnspath.[1] Mr. William Patterson, late of the Abbey Farm, North Berwick, who, having lived many years in Coldingham Parish, is well acquainted with St. Abb's Head and the neighbouring coast, has informed me that, about forty years ago, he harried the nest of a Raven near Dowlaw, a little to the west of Fast Castle. The nest was situated in a precipitous cliff overhanging the sea, and as he could not reach it by climbing. he tried to get the young ones out of it by letting down upon them a fleece of wool attached to the end of a long rope, in the expectation that they would clutch at the wool with their claws, and hang on to it until it was drawn by the rope up to the top of the precipice; but although they seized the wool with their feet, they always fell back into the nest on an attempt being made to draw up the fleece. When he saw that this plan was a failure, he procured a piece of paling about two feet long; and, attaching it by the middle to the end of the rope, he let it down to the nest, and jerked out the young birds, five in number, by pulling the rope sharply when they got on either end of the paling. On being thus thrown out of the nest, they fell down the rocks to the bottom of the cliff, and were all killed but one, which had apparently been a little better fledged than the others, for it fluttered down to the beach, where he secured it. He kept it for some years, and it became quite tame; but one day it attacked a little boy about four years old, and would have pecked out his eyes had he not protected them with his hands, on which account it was destroyed. The nest used to be near Fast Castle every year in those days. In a note dated

---

[1] *New Stat. Acc. of Scot.*, vol. ii., Berwickshire, p. 300.

1842, Mr. Hardy says that " a man named Bookless ventured over the rocks near Petticowick to harry the Raven's nest, the rope being held by two other weavers, George Pae and Robert Craig—both, 'na great rugs.' Bookless was a kind of ' Wandering Jew,' with no settlement in him, who would do anything to win a penny-piece rather than weave. The promised reward was five shillings from the farmer at Northfield, some of whose sheep the birds had destroyed by pecking out their eyes when lying on their backs. The rope, being old, broke on a sharp rock, and Bookless fell sheer down to the sea-beach, and was killed." Writing in 1858, Mr. Hardy mentions that " near the Rammel Cove there is a bald, steep rock facing the east, crowned with long heather ; and there the Raven used to build, and likewise the Peregrine Falcon. He also states that the former had its nest of old at the Swallow Craig, on the coast near Oldcambus, and in the rocks at Fairneyside, near Burnmouth. Mr. Wilson, late of Edington Mains, has told me that, when he was a boy, a pair of Ravens built their nest every year near the southern extremity of that farm, at a place called the " Blue Braes," on the banks of the Whitadder, a little distance above Hutton Hall Mill. They continued to do so until 1825 or 1826, when, in consequence of the depredations committed by the old birds amongst the young poultry, he assisted in harrying the last nest, which was placed on a tall ash-tree near a rock at the Braes. The young birds, which were full fledged, could be seen sitting in the nest, and were shot; while the old ones hovered high in the air out of reach. After this they never returned to build there. In his notes on the "Birds of Lauderdale," Mr. Kelly mentions that the Raven used to nest at West Hope.[1]

Mr. Cowe, Oldcastles, tells me that, in his boyhood,

---

[1] *Hist. Ber. Nat. Club*, vol. vii. p. 301.

between 1815 and 1830, he frequently saw sheep which had the misfortune to be "whammelled" on his father's farms of Blackpotts and East and West Muirside, near Coldingham, with their eyes pecked out by this bird.

Selby states that on the occasion of the visit of the Berwickshire Naturalists' Club to St. Abb's Head, on the 18th of July 1832, three Ravens were seen; which, after soaring around and attaining a considerable elevation, moved inland, uttering at intervals their loud and raucous croak.[1] Writing in 1851 on "Birds found at St. Abb's Head," Mr. Archibald Hepburn says:—"A solitary pair of Ravens on Weston Thirl Cliff, are the only representatives of a once numerous and daring band of plunderers. Many years ago their depredations on the young lambs were so serious as to render their destruction a matter of great importance to the flock-masters; these birds used to watch for the birth of a lamb, and, before the mother was able to attend to its safety, its eyes and tongue were pecked out, the umbilical cord was rudely seized, the intestines were thus uncoiled and dragged forth, and the little sufferer soon dropped to rise no more."[2] Mr. Patterson, Ancrum Woodhead, informs me that, about thirty-five years ago, he trapped a specimen on the Black Hill, near Earlston, using the leg of a rabbit for bait, and that the bird was stuffed for Mr. Cotesworth of Cowdenknowes. When the Berwickshire Naturalists' Club met at Coldingham on the 26th of May 1859, the members were conducted by Mr. Herriot of Northfield round the coast by St. Abb's Head, and on this occasion two pairs of Ravens were observed.[3] Dr. Stuart writes:—"From 1864 to 1868, at Earnsheugh, near St. Abb's Head, the highest rock precipice between Leith and London, they nested every year. They became a perfect nuisance to the sheep on the

---

[1] *Hist. Ber. Nat. Club*, vol. i. p. 19.
[2] *Ibid.* vol. iii. p. 72.
[3] *Ibid.* vol. iv. p. 131.

farm of Northfield, occupied by the late Mr. Herriot. No sooner did a ewe roll over on her back, and fail to right herself, than she was pounced upon by the Ravens, which pecked out her eyes. Ever on the alert, they never missed a chance, and many a poor sheep and lamb suffered from their audacity. The patience of the shepherds could stand their behaviour no longer; and Andrew Weatherston, the head man, hit upon a plan which effectually put a stop to their depredations. Having obtained a tar-barrel, and attached a chain and a long rope to it, he went to Earnsheugh at night. He had previously made himself acquainted with the exact situation of their nest in the inaccessible cliffs, and, getting his tar-barrel lighted, he hoisted it over the rocks to the nest, destroying the young, and frightening the old birds from the coast entirely for some time." [1] About 1873 two birds of this species were shot on Coldingham Moor by James Blythe, gamekeeper, Netherbyres, and were presented to the Museum of Science and Art in Edinburgh. Mr. Hardy records that on the 24th of July 1874, a young Raven appeared at Siccar Point, and that, on the 4th of August following, three young ones were seen frequenting that place.[2] In the previous year a specimen was caught in a rabbit trap in Edgarhope Wood, Lauderdale, where the bird is now looked upon as a rare visitor.[3] A Raven was seen at a dead horse near Oldcambus on the 23rd October 1876,[4] and two were noticed on the sea-banks near Burnmouth on the 4th of December following.[5] Mr. James Smith, shepherd, Byrecleugh, tells me that, about 1877, two were seen on Greencleugh Ridge in August, and one of them was shot by the Duke of Roxburghe. A Raven frequented Lamberton Moor in January 1879,[6] and about the same

---

[1] *Hist. Ber. Nat. Club*, vol. xi. p. 245.
[2] *Ibid.* vol. vii. p. 281. [3] *Ibid.* vol. vii. p. 303.
[4] Mr. Hardy's *MS. Notes*.
[5] *Hist. Ber. Nat. Club*, vol. viii. p. 179. [6] *Ibid.* vol. viii. p. 505.

time one was noticed at Siccar Point.[1] The last-mentioned locality seems to be a favourite resort, for they were observed there on the 24th January of 1880,[2] and the 29th of January 1882.[3] The Rev. George Cook, of Longformacus, writes to me that a Raven was seen near Whitchester in the spring of 1883.

This species is omnivorous, but its principal food consists of carrion, such as the carcases of horses and sheep; and includes dead fish and other marine creatures, which it finds on the sea-shore. It also preys upon small birds and quadrupeds, and will attack larger animals when they show signs of weakness—as in the case of a sheep lying on its back and unable to rise, or a young lamb—its first proceeding in this case being to peck out the eye, and then fall upon the intestines.

It was not a very early breeder with us, generally beginning to build in March, and having eggs in April. The nest, which is usually constructed of sticks lined with wool, is placed in some steep cliff overhanging the sea, such as may be seen at St. Abb's Head; though formerly it appears not only to have occupied such positions in this county, but also to have been built in inland precipices along the courses of the Whitadder and Blackadder, and in the Lammermuirs, some parts of which have been already mentioned as still bearing the name of the bird.

The eggs, which vary from four to six in number, are bluish green blotched and streaked with dark olive-brown.

The Raven when taken young is easily tamed, and, with a little trouble, may be taught to speak. A fine bird in beautiful plumage has for many years been in the possession of Mr. John Miller, now forester at the Grange, near Coldingham, which can imitate the human voice so well that a

---

[1] *Hist. Ber. Nat. Club.* vol. viii. p. 527.
[2] *Ibid.* vol. ix. p. 389.
[3] *Ibid.* vol. ix. p. 556.

listener, when the bird is out of sight, is very apt to be deceived. This peculiarity of the Raven is referred to by Sir David Lindsay in his *Complaynt of Scotland*, where he says :—" He be grit subtiltie neurissit tua zong Corbies in tua cagis in tua syndry housis, and he leyrnit them bayth to speik."

THE SKYLARK.

LAVEROCK, FIELD LARK, COMMON LARK.

Alauda arvensis.

𝕿𝖍𝖊 𝕷𝖆𝖛𝖊𝖗𝖔𝖈𝖐.

Hark! hark! the Lark at heaven's gate sings,
  And Phœbus 'gins arise,
His steeds to water at those springs
  On chaliced flowers that lies.
          SHAKESPEARE, *Cymbeline*, ii. 3.

The Lauerock maid melody up hie in the skyis.
          SIR DAVID LINDSAY OF THE MOUNT.

THE joyous song of this charming bird is heard in the Merse from early spring until summer is well advanced, and during that period, from the grey dawn of the morning until the shades of the evening, he pours his gladsome music from the sky.

> Up in the morning while the dew
>   Is splashing in crystals o'er him;
> The ploughman hies to the upland rise,
>   But the lark is there before him;
> He sings when the team is linked to the share,
> He sings when the mist is going,
> He sings when the noontide south is fair,
> He sings when the west is glowing.
>   COOK.

The Skylark generally rises from the ground on fluttering wings as it commences its song, soaring upwards singing until it is almost lost in the sky; and when it has attained its highest flight it comes warbling down to join its mate.

> Slow the descent at first, then by degrees
> Quick and more quick, till suddenly the note
> Ceases; and like an arrow-fledge he darts,
> And, softly lighting, perches by her side.
> GRAHAME.

It sometimes, however, sings when on the ground. Mr. Archibald Hepburn, Whittinghame, East-Lothian, observed that the duration of each song in early spring was about two or three minutes, and, during summer, not longer than a quarter of an hour at the most. It is occasionally heard singing in September and October, but the notes are then generally more broken than in spring and summer. The association of the song of the Lark with happiness and freedom in this county in olden times, is referred to by David Hume of Godscroft, in his *History of the Douglases*, who says, with reference to Archibald, sixth Earl of Angus, son of "Archibald Bell-the-Cat," that during the siege of Tantallon Castle, in 1528, "the Earle himselfe remained at Billie in the Merse, within his barony of Bonkle, not willing to shut himself up within the wals of any strength, having ever in his mouth this maxime (which he had received from his predecessours) that it was better to hear the Lark sing than the mouse cheep."[1] When a person is heard whistling joyously in the county at the present day, a popular remark sometimes made is, that he is "whistling like a Laverock."

The Skylark is a partial migrant in Berwickshire, and although some remain with us throughout the year, yet comparatively few stay over the winter; the great majority departing towards the south in autumn, and returning in spring. In summer it is found in pairs more or less throughout

---

[1] *History of the House and Race of Douglas and Angus*, by David Hume of Godscroft, p. 259.

all the county; but in some localities, such as the neighbourhood of Gordon,[1] it is more plentiful than in others. It is by no means numerous in the vicinity of Paxton, and possibly the large rookery here may have something to do with its scarcity, for it has been alleged that Rooks destroy the eggs of Skylarks.[2] Towards the end of harvest they are often seen by the partridge-shooter in stubble and grass fields, and are frequently a source of annoyance to him, on account of his dogs making false points at them, their scent being apparently somewhat similar to that of the Partridge. They are then generally found singly, and lie very close to the dog's point; but, as autumn advances, they become gregarious, assembling in considerable numbers and feeding on the stubble fields. Enormous flocks of Skylarks are seen on migration at the lighthouses on the east coasts of England and Scotland in autumn and spring.[3] In severe snow-storms, such as those of the winters of 1878-79 and of 1879-80, great numbers sometimes visit the neighbourhood of the sea-coast, when the snow is not lying so thickly there as in the more inland districts. On the 27th of January 1879, during the continuance of the heavy snow-storm, which was then lying all over the interior

---

[1] *Hist. Ber. Nat. Club*, vol. ix. pp. 229, 235.

[2] A writer in the *Scotsman*, dating from Orkney, 26th April 1886, says, that in Shetland, where there are no Rooks, Skylarks are very plentiful, and that in Orkney Larks were likewise very numerous until about ten years ago, when Rooks first came to that island. Since then the Rooks have increased largely, and the decrease in the number of Skylarks is marked.

[3] Flocks of Skylarks were seen on migration in *Autumn* 1880.—At Farnes, 30th Sep. to 31st Oct. *Autumn* 1881.—Isle of May, 24th to 27th Sep. *Spring* 1882.—Isle of May, 15th March. *Autumn* 1882.—Isle of May, 11th to 18th Sep. and in October. Mr. Cordeaux records immense flocks crossing east coast of England both day and night, at twenty-nine stations. *Spring* 1883.—Isle of May, 11th Feb. *Autumn* 1883.—Isle of May, 21st Sep. to 1st Nov., Mr. Cordeaux records immense numbers at all stations on the east coast of England. *Autumn* 1884.—Isle of May, 12th to 16th Nov., great flocks; also along whole of east coast of England, from 6th Sep. to 23rd Dec. *Spring* 1885.—Isle of May, 18th Feb.; Farnes, 7th April. *Autumn* 1885.—Isle of May, 14th Oct., great flocks; Farnes, 12th to 20th Oct.—*Reports on Migration of Birds*, 1879-85.

of the county, I saw large flocks on Lamberton Farm, situate on the coast between Burnmouth and Berwick-on-Tweed, there being comparatively little snow on the ground facing the sea there. Immediately previous to the two terrible drifty days, the 1st and 2nd of March 1886, immense flocks, consisting of many thousands, were observed in the neighbourhood of Mordington; and, ten days later, I saw very large numbers on the farm of Paxton South Mains. The Skylark prefers an open exposed district away from the shelter of woods, and often roosts at night on the barest places. This characteristic is alluded to in a remark which, according to the late Dr. Henderson, of Chirnside, used to be often heard amongst the peasantry of Berwickshire. On a cold stormy night, when the cottage smoked and the goodwife complained of the inconvenience, it was usual for the head of the family to say: "We'd better hae reek than cauld; there's nae reek in the Laverock's house the nicht."[1] A Scottish weather proverb says: "As lang's the Laverock sings afore Candlemas, it greets after't," which apparently alludes to the popular belief that severe weather often follows a mild January.

This bird has been long known as a very savoury morsel for the table, and roasted Larks seem to have been a favourite dish both in England and Scotland in olden times.

> The wyfe said, speid the kaill cir soddin,
> And als the Laverok is fust and loddin.[2]
> *Bannatyne Poems.*

> There is a well-fair abbéy
> Of white monkés and of grey,
> There beth bowers and halls;
> All of pasties beth the walls.
> Of flesh, of fish, and a rich meat
> The likefullest that man may eat.

---

[1] Dr. Henderson's *MS. Notes.*
[2] "Roasted and swollen," Jamieson's *Scot. Dict.*

> The Leverokes that beth couth,[1]
> Lieth adown to man-is mouth
> Y-dight in stew full swirthe[2] well,
> Powder'd with gingelòfre and canell.[3]
>
> *Land of Cokaine.*

At the present time as many as twenty or thirty thousand Skylarks occasionally reach the London market in a single day, and it has been estimated that about £2000 worth are annually sold in the metropolis alone.[4] They are mostly caught by dragging nets at night over stubbles and fallows.

A peculiar feature in the Skylark is its very long hind claw.[5] This is referred to in the following lines, which are thought to resemble its song,—

> Up in the lift go we,
> Te-hee, te-hee, te-hee, te-hee!
> There's not a shoemaker on the earth
> Can make a shoe to me, to me!
> Why so, why so, why so?
> Because my heel is as long as my toe![6]

Although the plumage of this bird seldom varies, yet occasionally we have instances of albinos in the county. The Earl of Home's gamekeeper observed a cream-coloured specimen on Drakemire Moor in 1875,[7] and Mr. J. Waddell found a white example in a turnip field near Birgham a few years ago.

They are very fond of dusting themselves in dry sunny weather, and are frequently seen doing this in the middle of our public roads in summer.

The food consists of seeds—including corn—insects, and worms. The nest is built on the ground amongst herbage,

---

[1] Taught. [2] Quickly. [3] Ginger and cinnamon.
[4] See *The Standard*, 16th Jan. 1886.
[5] An interesting notice of the long heel-claw of the Skylark, by Mr. Ralph Carr, of Hedgeley, appears in the *History of the Berwickshire Naturalists' Club*, vol. iv. p. 209.
[6] *Popular Rhymes of Scotland*, R. Chambers.
[7] *Hist. Ber. Nat. Club*, vol. vii. p. 512.

THE SKYLARK.

and is composed of dry grass, with a few rootlets. The eggs, which are four or five in number, are generally dull white, so thickly freckled and spotted with olive-brown as almost to conceal the ground colour.

The following places in Berwickshire appear to have derived their names from this bird :—Laverock Law, between Fogo village and Charterhall; and Laverock Law,—high ground to the east of Moorhouse in Coldingham parish, where there is also a small stream called Laverock Law Burn.

THE SWIFT.

BLACK MARTIN, SCREECH, DEVILING.

Cypselus apus.

𝕮𝖍𝖊 𝕮𝖗𝖆𝖓.

To mark the Swift in rapid giddy ring
Rush round the steeple, unsubdued of wing.
                          GILBERT WHITE.

THE Swift is later in coming to the county than the three species of Swallows; not usually making its appearance until about the second or third week of May, when it may be seen frequenting its accustomed haunts, such as old castles and towers, steeples, and other lofty buildings, as well as old houses in villages and towns. Amongst its favourite resorts may be mentioned Hume Castle; Cockburnspath Tower; Evelaw Tower; and Dryburgh Abbey. At the last-mentioned place especially, it may be seen in great numbers; wheeling aloft on the wing, and from time to time rushing round the beautiful ruins of the old monastery, uttering its harsh screams, and entering the holes in the mouldering walls where it nests.

    This bird comes to us from Africa late in spring, and migrates southwards again in August or early in September. It constructs a slight nest of pieces of straw and a few feathers cemented together with its viscid saliva, placed in a hole in a wall or under the tiles of an old house; and the eggs, which are generally two in number, are white. Only one brood is reared in the season. The food of the Swift consists entirely of insects.

THE NIGHTJAR.

GOATSUCKER, FERN OWL, CHURN OWL, JAR OWL, DOR HAWK, NIGHT HAWK, NIGHT CHURR.

Caprimulgus europæus.

The Night Hawk.

While o'er the cliff th' awaken'd Churn Owl hung,
Through the still gloom protracts its chattering song.

GILBERT WHITE.

THE arrival of this interesting bird in the county does not generally take place until May is well advanced, it being one of the latest of our summer migrants. Shortly afterwards it may be seen in the twilight of fine evenings, hawking for moths and other insects along the borders and glades of the woods, and wheeling round trees, in its usual haunts, which are secluded wooded districts in the vicinity of moors, where ferns abound; the neighbourhood of Abbey St. Bathans[1] and Penmanshiel being favourite resorts. Here it may be observed following a regular line of flight many times in succession; if the wind or weather change, its beat is altered to places where its prey is most plentiful. The Nightjar being a nocturnal bird is, however, oftener heard than seen, the peculiar churring note which it utters in the dusk of the evening, while resting on

---

[1] Mr. Kelly records that Nightjars are plentiful about Abbey St. Bathans, where he says he has occasionally seen them hunting like Swallows for flies amongst cattle.—*Hist. Ber. Nat. Club*, vol. viii. p. 145.

a tree or other elevated perch, striking the ear at a considerable distance. Writing in 1834, the Rev. Andrew Baird, minister of the united parishes of Cockburnspath and Oldcambus, says that the Nightjar has been frequently shot in that locality.[1] Mr. Hardy mentions, under date the 17th of June 1837, that its " curr," as heard by him on a woodland road overshadowed by oaks in Penmanshiel Wood, affected the air for a considerable distance, and, when he was within four yards of the bird, "the movement of the pulsations of the air within the ear was not very pleasant." On the 27th of July 1839, he notes that " in Birchy Bank, which is part of the wood below Penmanshiel, lying above the railway tunnel, I surprised a Goatsucker on her nest. She sat still a long time although I could have touched her, but on my putting forth my hand she fled with a loud scream, fluttering in a zigzag manner like a wounded bird. In the nest, which was merely a hole scraped in the mossy soil, with a number of pismires running about in it, were two callow young ; one clothed with down, the other, which was half out of the egg, not showing any covering."[2] The nest seems to have been frequently discovered about Penmanshiel in those days, for I find Mr. Hardy again recording, in 1842, that " A Nightjar had its nest in one of our fields called Cuddy's Stele, near Red Clews Cleugh." Blackburn Rigg, Ewieside, and Dowlaw Dean are mentioned by him as favourite haunts of this bird, and he speaks of hearing its " chuck, chuck " call in the woods there in the gloaming.[3] Mr. W. Duns, Duns, has informed me that about thirty years ago the Nightjar used to be rather common about Longformacus,[4] and at Buchan's Moor on Cumledge Estate, also at Barrow Mill ; and that

---

[1] *New Stat. Acc. of Scotland*, vol. ii., Berwickshire, p. 291.
[2] Mr. Hardy's *MS. Notes*.   [3] *Ibid.*
[4] A specimen shot at Longformacus was exhibited at the meeting of the Berwickshire Naturalists' Club held at Duns on 27th June 1867.—*Hist. Ber. Nat. Club*, vol. v. p. 307.

one was shot at Abbey St. Bathans by Shearlaw, the gamekeeper there, about 1874. Mr. Lockie writes, 1887, that one was obtained at Spottiswoode several years since, and Mr. Charles Watson, Duns, informs me that a specimen was procured at Cockburn in 1873, as late in the season as September. The Nightjar is very rare in the neighbourhood of Paxton. I have seen it here only once during the last seventeen years—on the 19th of May 1878. The gamekeeper at Mordington tells me that it is sometimes observed about the "Lang Belt" plantation there; and Mr. John Thomson reports to me that it is a regular summer visitor to the neighbourhood of Mertoun and Cowdenknowes. Mr. Pringle, Ayton Castle, has informed me that he shot a Nightjar on Cocklaw Farm on the 10th of September 1887, and that one was killed at The Press about the same time.

The bird usually roosts on the ground amongst ferns or other herbage; and on being disturbed it generally flies to a large tree, if one be near, crouching down along one of the branches, and not sitting across it, in the usual manner of birds. Its food consists solely of insects, such as large moths and night-flying beetles. The eggs, which are two in number, are laid in June, upon the bare ground—often amongst brackens in the vicinity of woods—and are very beautiful. Their texture is smooth, and their colour white, mottled, clouded, blotched, or spotted with lilac, grey, and brown of various shades.

THE GREAT SPOTTED WOODPECKER.

PIED WOODPECKER, WITWALL, WOOD PIE, FRENCH PIE.

Dendrocopus major.

> *The hazel blooms in threads of crimson hue,*
> *Peep through the swelling buds, foretelling Spring,*
> *Ere yet a whitethorn leaf appears in view,*
> *Or March finds Throstles pleased enough to sing.*
> *To the old touch-wood tree Woodpeckers cling*
> *A moment, and their harsh-toned notes renew.*
>
> CLARE, *The First Sight of Spring.*

THE first record which we have of the occurrence of the Great Spotted Woodpecker in Berwickshire is in the *Old Statistical Account of Scotland,* where the Rev. Thomas Mills, in his report on the parish of Ladykirk, written in 1793, remarks that, "in winter Woodpeckers sometimes appear."[1] Dr. Stuart of Chirnside contributed some admirable notes on this Woodpecker to the *History of the Berwickshire Naturalists' Club* for 1868, in which he says:—"Previous to the meeting of the Club at Chirnside,[2] Dr. Maclagan of Berwick wrote to me an interesting letter relative to *Picus major.* When resident in Canada, he was familiar with the habits of this bird, and well acquainted with the peculiar tapping noise it makes when searching for insects. He was therefore not a little astonished when walking near Berwick to hear the familiar sound, and pre-

---

[1] *Old Stat. Acc. of Scotland,* vol. viii. p. 74.
[2] The meeting referred to was held on 26th September 1868.

sently a fine Woodpecker commenced operations on a rail, close to where he was standing; so he watched and satisfied himself as to its identity. On regaining the road, he was overtaken by Mr. Smith, residing at Letham, who informed him that he had a rare bird in his pocket; upon producing which, strange to say, here was another specimen of *Picus major*, shot at Monynut in the Lammermuirs. Upon visiting the bird-stuffer in Berwick, he was shown three other specimens, one of which had been shot near Eyemouth. A few years ago Lord William Kennedy shot a Great Spotted Woodpecker in Edington Hill Wood, about a mile from Chirnside; and Mr. Stewart, at one time residing at Blanerne, shot another most beautiful specimen on Leaderside, in this county. The first of these two examples was certainly killed in the shooting season, and, if my memory is right, about the month of November or December."[1] Mr. John Aitchison, plasterer, Duns, has told me that a bird of this species was shot at Channobank, near Abbey St. Bathans about 1850; and Mr. Scott, Lauder, has informed me that one was killed in Edgarhope Wood, Lauderdale, about 1860. In September 1868, when this bird appears to have been seen in considerable numbers in various parts of Scotland—including Berwickshire—Mr. Compton-Lundie shot one at Spital House; and about the same time the forester at Paxton House saw another climbing up a larch tree in the policy-ground there. On the 20th of September of that year a Great Spotted Woodpecker was seen at Mordington, in the gamekeeper's garden. In 1874, Miss Georgina Milne-Home presented to Berwick Museum a specimen which was got at Milne Graden a few years before that date. Mr. Cowe, Lochton, tells me that two were shot in a wood by the side of the Tweed there in 1874. Writing in

---

[1] *Hist. Ber. Nat. Club*, vol. v. pp. 409, 410.

1875, Mr. Hardy mentions that in summer one was seen clinging to a paling on the coast near Dunglass;[1] and, in 1876, he says that three were seen in Penmanshiel Wood some years before that date, one of them being observed near the Tower Farm.[2] According to Mr. Robert Waite, Blinkbonnie, Duns, a Great Spotted Woodpecker was shot at Threeburn Grange, near Coldingham, in November 1876.[3] Mr. George Fortune, architect, Duns, says that he saw one on a tree in a plantation by the side of the public road near Longformacus, some years ago; his attention having been drawn to the bird by the noise which it made in striking the tree with its bill. A female was killed at the White Gate on Blackerston estate, near The Retreat, in January 1879, and is now in the possession of Mr. Hogg of Quixwood. The gamekeeper at Abbey St. Bathans states that he saw a specimen on a Scotch fir in Lippie Plantation, in November 1886, having heard it tapping with its bill at a distance of forty yards; and Mr. Pringle, Ayton Castle, tells me that his gamekeeper shot a male at the kennels there, on the 20th of November 1887; while on the 2nd of January 1888, Mr. Mitchell-Innes of Ayton was so good as to send to me a beautiful specimen of the female, which was shot by him near Ayton Castle, shortly before that date.

The Great Spotted Woodpecker is an irregular autumn visitor to the county, being seen in small numbers in some years, such as 1868, whilst many years sometimes elapse without any being observed. It comes with the migratory flocks of birds which visit our shores from the northern parts of Europe at the season mentioned, and it has been occasionally observed on migration at the lighthouses on the coasts of Scotland and England.[4]

---

[1] *Hist. Ber. Nat. Club*, vol. vii. p. 514.   [2] *Ibid.* vol. viii. p. 193.
[3] *Ibid.* vol. viii. p. 196.
[4] See *Reports on the Migration of Birds*, 1879-85.

It is of a solitary and wary disposition, and chiefly haunts secluded woods and plantations, where the trees are old and large. There it may be seen climbing the trunks and higher branches in search of the insects concealed in the crevices of the bark, and may be heard at a considerable distance tapping on the trees with its bill. It likewise feeds on nuts, acorns, the seeds of the fir-tree, and various berries, including those of the mountain ash. The Great Spotted Woodpecker is about the size of a Song Thrush, and may be easily identified, if seen in our woods, by its conspicuous plumage; the back being glossy black, each wing having a large white spot upon it, and the hind-part of the head of the male being crimson. The under-parts are dirty white, and the vent and lower tail-coverts also crimson. It has not been known to breed in Berwickshire.

THE WRYNECK.

CUCKOO'S LEADER, CUCKOO'S MATE, CUCKOO'S MESSENGER, PEA BIRD, SUMMER BIRD, EMMET HUNTER.

Iynx torquilla.

> *Full nature swarms with life;*
> *The flowery leaf*
> *Wants not its soft inhabitants. Secure*
> *Within its winding citadel, the stone*
> *Holds multitudes. But chief the forest boughs*
> *That dance unnumber'd to the playful breeze,*
> *The downy orchard and the melting pulp*
> *Of mellow fruit, the nameless nations feed*
> *Of evanescent insects.*
>   THOMSON, *Summer.*

ALTHOUGH the Wryneck is a common summer visitor to the south-east of England, and is occasionally seen in the northern counties, it has been very rarely observed in Berwickshire. There is no record of its occurrence in the county, in the *History of the Berwickshire Naturalists' Club*—a publication which, for the last fifty-four years, has been devoted to the natural history of Berwickshire and the surrounding district.[1]

Colonel Milne-Home of Wedderburn has informed me that on the afternoon of the 31st of July 1887, he and his sisters, the Misses Milne-Home of Milne Graden, saw a Wryneck on the stone side-post of a window on the principal staircase of Milne Graden House. When it was first

---

[1] It is stated in Yarrell's *British Birds* (vol. ii. p. 156, 1st ed.) that "there are records of this bird [the Wryneck] having been killed twice in Berwickshire," but no locality in the county is mentioned.

noticed the bird was sitting in a perpendicular position, with its head upwards, clinging to the stones with its claws; and shortly afterwards it began to twist its head and neck about in a very peculiar manner until they turned round half a circle, while the lower part of the body, claws and tail, remained quite unmoved. Ultimately it slipped round the gable of the house, and flew towards the top of a neighbouring pine-tree, where it went out of sight.

The Wryneck is a beautiful bird of an elegant form, about the size of a Sky-lark, and its colours are soft and varied, like those of the Nightjar and Owl. In its habits, which are shy and retiring, it somewhat resembles the woodpeckers; and, like those birds, it searches the trunks and branches of trees for insects. It is very fond of ants and their eggs, and is often seen on the ground about their hills. The tongue of the bird, which is covered with a glutinous substance, is used with great effect amongst ants' eggs, for when it is protruded they adhere to it, and are thus rapidly conveyed to the mouth. This species generally arrives in England about the first or second week of April, and on account of its usually preceding the Cuckoo by a few days, is called in some districts the Cuckoo's Leader, the Cuckoo's Mate, or the Cuckoo's Messenger.

THE KINGFISHER.

Alcedo ispida.

> *There came,*
> *Swift as the meteor's shining flame,*
> *A Kingfisher from out the brake,*
> *And almost seemed to leave a wake*
> *Of brilliant hues behind.*
> FABER.

THE Kingfisher is found on all our streams, where its brilliant plumage is never seen to greater advantage than when it glances in the sun, as the bird, darting from some retreat by the water-side, rapidly wings its way along the course of the river. On the Tweed it is occasionally observed at Paxton, and more frequently higher up the river in the neighbourhood of Ladykirk,[1] Milne Graden, Coldstream, Lees, Mertoun, and Gladswood. Mr. Thomas Hood, Coldstream, has informed me that there was a Kingfisher's nest near Lees in the summer of 1886; and Mr. John Fulton, salmon-fisher, says that a pair bred near Damford Shiel, at Milne Graden, two or three years ago.

It is an early breeder, and the eggs are asserted to have been taken on the Tweed, for the second time in the season, from the same pair of birds, by the 10th of April.[2] The Whitadder seems to be the favourite resort of the Kingfisher in Berwickshire, and the nest has frequently been found on the

---

[1] Mr. John Blair, artist, has mentioned to me that he often saw Kingfishers at the side of the Tweed at Ladykirk in the summer of 1887, where he believes they nested that season.

[2] *Hist. Ber. Nat. Club*, vol. vii. p. 285.

banks of this stream and several of its tributaries. Mr. John Ferguson, Duns, writes:—" In July 1872, I discovered two nests within a hundred yards of each other, in a sand-bank at the side of a burn (the Cabby Burn) which flows into the Whitadder a little above Hutton Hall. What attracted my attention to them was the quantity of excrement running out of the holes in which they were placed. These holes were bored, in the manner of a Sand Martin's nest, into the bank, and were from two and a half feet to a yard in length. On inserting my hand, I found three young birds in one nest and seven in the other, all nearly full feathered. They were of the same colour as the old birds, but the whitish bar on the shoulder was scarcely so distinct. Quantities of small fish bones, which had apparently been ejected from the stomachs of the birds, were lying strewed about at the farther end of the hole, and on these the young were sitting. The nests themselves were quite clean, although the entrances were so choked up with filth that it was a marvel the old birds could get in. I concealed myself behind a hedge which runs parallel to the course of the burn, within a few feet of one of the nests, and, after waiting a few minutes, saw one of the parent birds, carrying a minnow in its bill, fly rapidly up and down the burn several times, and finally alight on a stone immediately beneath the hole. The minnow was carried by the tail and was evidently quite dead. After looking about suspiciously for a little, the bird entered the nest. It remained there about a minute, and then darted out with the speed of an arrow. The manner in which it left the nest contrasted strangely with its cautious mode of approaching it. I once discovered a Kingfisher's nest with eggs. These were of a pinkish white colour, which became a pure white when they were blown. The only notes I ever heard the Kingfisher utter are a sort of 'chuck,' and a peculiar indescribable sound resembling that made by a dog when retching. When emitting the latter note the bird contracts the fore-part of its body with a jerking motion, and, as I never heard this cry except in summer when two birds were together, I concluded it was an amorous call, like that of the Rook in early spring."[1] Mr. John Gillies, Edington Mill, says that it is often seen on the river there, and at Ninewells; and Mr. John Blackadder, East Blanerne, has mentioned to me that he observes it about Billie Burn every spring. There is a deep pool on the Whitadder, near Elbaw, called "Hell's Cradle,"[2] and here, I have been told by the gardener at Abbey St. Bathans, a nest of this species, with young, was found in the summer of 1886. Mr. Duns, Duns, has known it build near the Peelie Braes, between Cockburn Mill and Preston Bridge, and the nest has also been discovered in other parts of the river. On the Leader, and the streamlets which flow into it, the Kingfisher is occasionally seen, and Mr. Scott, Lauder, told me that it had bred at Carolside within the last few years. Mr. Kelly writes:—"A pair for a long time frequented the Longcroft Water, from where it joins the Leader to Cleekhimin, and nested in the Red Brae. There being no trees, they were obliged to watch their prey from a rock."[3] The Boondreigh Burn, which is another tributary of the Leader, sometimes receives a visit from this bird, no fewer than four having been seen near the Dod Mill, a few years ago.[4] Mr. Hardy frequently notices it on the Eye about Renton and Grantshouse, and Mr. J. L. Mack tells me that he found a nest with four eggs in a hole

---

[1] *Hist. Ber. Nat. Club*, vol. vii. p. 120.
[2] In this pool were found the bodies of two girls—Isabella Lauder, Cranshaws, and Matilda Whitehead, Duns—who were unfortunately drowned on 30th September 1883, while attempting to cross the Whitadder on a foot plank at Ellemford. Their bodies were not recovered for some days after the accident, having been carried down to Elbaw by the flooded river.
[3] *Hist. Ber. Nat. Club*, vol. vii. p. 304.
[4] Mr. Lockie, Gateside, Spottiswoode.

in the bank of that stream near Coveyheugh House, in April 1878. Mr. Mitchell-Innes of Ayton has informed me that he often notices the Kingfisher on the Eye, between Ayton Castle and Millbank. It appears to be scarce on the Dye in the neighbourhood of Longformacus, according to Colonel Brown; and the Rev. George Cook tells me that he has seen it only twice on that stream within the last fifteen years, both instances being at the junction of the Dye with the Watch. It seems to visit the Blackadder in the neighbourhood of Greenlaw very seldom.[1] The lake at Mellerstain appears to be rather a favourite haunt;[2] and the bird has been twice noticed on the sea-coast near Burnmouth within the last few years.[3]

In some seasons a scarcity of this species has been observed on our rivers and streams, the cause of which is not known; but it is conjectured that it may be owing to sudden floods drowning the birds in their nests, or a severe winter, when the streams are all frozen over, forcing many of them to leave the district.

The food of the Kingfisher consists of small fish, such as minnows, and of aquatic insects. While watching for its prey, it sits on an overhanging branch, or on a rock or stone, whence it drops perpendicularly into the water, and, after remaining underneath for a few seconds, rises with its prey in its bill to its usual perch. Here, it kills the prey, if a fish, by striking it against the bough or rock, preparatory to swallowing it. It sometimes poises itself in the air like a kestrel for a few seconds, before making the descent into the water.

---

[1] Mr. Robert Renton, Greenlaw.
[2] Mr. James M'Pherson, gamekeeper, Mellerstain.
[3] Mr. Crockett, salmon-fisher, Burnmouth.

THE HOOPOE.

Upupa epops.

The green Cicada chirping 'mid the grass,
The crested Hoopoes singing as they pass.
MITCHELL, *Ruin of many Lands.*

THIS interesting bird is very rarely seen in Berwickshire. In the *Old Statistical Account of Scotland* (1795) it is stated that, "On 18th Sept. 1790, was found, three miles south-east from Duns, a bird very rare in Scotland. It was killed by a cat, and discovered to be the bird called Hoopoe by the English, Wedhope by the Germans, the Upupa of the ancients, and described by Pliny, Aristotle, Pausanias, Ælian, and others."[1] The Rev. A. Baird, writing in the *New Statistical Account of Scotland* (1835), mentions this bird as an occasional visitor to the parish of Cockburnspath.[2] In April 1884, Mr. Alexander Blackhall, Starchhouse Toll, near Mordington, showed me an old stuffed specimen very much decayed, which he said had been shot near Lamb's Mill, on the Whitadder, in a field of turnip seed, in July, about 1844. One was shot on Eyemouth Fort,[3] about the beginning of May 1879, and sent to a bird-

---

[1] *Old Stat. Acc. of Scot.* vol. iv. p. 392.
[2] *New Stat. Acc. of Scot.* vol. ii., Berwickshire, p. 299.
[3] The beautiful bay of Eyemouth is bounded on the north by a bold and picturesque headland, on whose summit the Duke of Somerset erected a fort in the year 1547, which was soon afterwards dismantled. Mary of Lorraine, during her regency, repaired the fortifications, which were, after the lapse of a few years, again demolished, and never afterwards repaired.

stuffer at Berwick for preservation, in whose shop it was seen by Mr. George Bolam. The gamekeeper at Mordington, Mr. James Purves, has informed me that one day in September 1883, when he and his son were going to Ayton, they saw a Hoopoe alight on the dyke between the moor and the Camp Field on Lamberton Farm. It put up its crest several times, and then flew away, and alighted amongst some burned whins on the racecourse.

I have never had an opportunity of observing the Hoopoe in this country, but have seen numbers at La Buisson Lusas, Loire et Cher, France, where they frequented the edges of woods and old orchards, and were very shy and wary. They built their nests in holes in trees, and were not very particular with regard to the materials which they used, as described in the following old French lines :—

"Dedans un creux, avec fange et ordure,
La Huppe fait ses œufs et sa maison." [1]

---

[1] See *The Folk-lore of British Birds*, Swainson, p. 106.

THE CUCKOO.

Cuculus canorus.

The Gowk.

Thrice welcome, darling of the Spring!
  Even yet thou art to me
No bird, but an invisible thing,
  A voice, a mystery;
The same whom in my school-boy days
  I listened to; that cry
Which made me look a thousand ways
  In bush, and tree, and sky.
To seek thee did I often rove
  Through woods and on the green;
And thou wert still a hope, a love;
  Still longed for, never seen.
And I can listen to thee yet;
  Can lie upon the plain
And listen, till I do beget
  That golden time again.
                    WORDSWORTH.

WHEN the woods of the county are preparing to put on their new robes, and the pale primrose is painting the deans, the voice of the wandering Cuckoo is usually heard for the first time in the season; and we hail the welcome notes, for they are associated in our memories with the joyous days of youth and early summer.

  Delightful visitant! with thee
    I hail the time of flow'rs,
  When heav'n is fill'd with music sweet
    Of birds among the bow'rs.

Michael Bruce, in his beautiful "Ode to the Cuckoo," from

which the above lines are taken, carries us back, like Wordsworth, to the days of our boyhood, and recalls " that golden time again"; and Montgomery makes the Cuckoo tell us why we love to hear its voice—

> Why art thou always welcome, lonely bird?
> —The heart grows young again when I am heard;
> Nor in my double note the magic lies,
> But in the fields and woods, the streams and skies.

From records of the arrival of this well-known spring visitor in Berwickshire for the last sixty-seven years, kindly furnished by Mr. Hardy, I find that its earliest appearance, during that period, was on the 4th of April 1833; whilst the latest date at which its notes were heard for the first time in the season was the 27th of May 1872, its advent being most frequently recorded on the 2nd of the latter month. The males arrive a few days before the females.

The Cuckoo is much more plentiful in the upland districts of the county, and along the edges of the Lammermuirs, than it is in the Merse; certain of the former localities being favourite resorts, such as Abbey St. Bathans, Ordweil, Wedderlie, Ecklaw, and Ewieside, where, about the beginning of May, as many as six or seven birds are sometimes in view at one time. In the moorlands the Meadow Pipit, Titlin, or Moss Cheeper (*Anthus pratensis*), is often seen following the Cuckoo, and on this account the proverb, " As grit as the Gowk and the Titlin " is occasionally heard amongst the country people. Sir David Lindsay, referring to that remarkable habit of the little bird, says, in his *Complaynt of Scotland*, " The Titlene follouit the Goilk and gart hyr sing Guk, guk." The motive for this action is not known; but it is supposed by some of our leading ornithologists to proceed from animosity, in the same way as Swallows follow a hawk.

In Berwickshire this species generally lays in nests of the Titlin, one egg only being placed in each. Mr. Hardy relates that, in the summer of 1842, he discovered near Red Clues a nest, in which, along with three Titlin's eggs, was one of the Cuckoo, the latter not being much larger than the former. The Titlin hatched the strange egg and two of her own, the third being unproductive; but soon afterwards the young Cuckoo threw its two companions and the addled egg out of the nest, near which Mr. Hardy found them lying when he next visited the spot. A full account of the manner in which the intruder accomplishes this feat is given in Yarrell's *History of British Birds*.[1] When the nest of the intended foster-parent is situated close under a bank, or in such a position that the Cuckoo cannot lay in it in the ordinary way, she has been observed to deposit her egg on the ground hard by, take it up in her bill, and place it in the nest. The egg of the Cuckoo is generally of a pale greyish green, or reddish grey, more or less closely mottled with darker markings, but it varies to bright blue and other colours. Two in my collection, which were found in Titlins' nests, are very like the eggs of that bird, but somewhat larger. When the young assassin has ejected the rightful occupants, it monopolises the whole attention of the foster-parents, grows rapidly, and leaves the nest in about a fortnight. For some weeks afterwards it continues to be fed by them, and they have sometimes been observed alighting on its shoulders and placing the food which they have collected in its gaping mouth.[2]

The young Cuckoo, when fully fledged, differs greatly in

---

[1] Yarrell, *History of British Birds*, 4th ed., vol. ii. p. 395.
[2] This habit is mentioned by Yarrell in his *History of British Birds*. Mr. Kelly states (*Hist. Ber. Nat. Club*, vol. vii. p. 522) that the late Mr. Simson, Lauder, when living at Edgarhope, saw a Meadow Pipit feeding a young Cuckoo by alighting on its shoulders and placing the food in its mouth.

the general colour of its plumage from the old bird,—being clove-brown, barred with reddish brown,—and at this stage it is sometimes shot by mistake for a hawk. The old Cuckoos leave the county in the beginning of July, but the young ones stay with us until August,[1] and sometimes the beginning of September.[2]

The food appears to consist of various insects and caterpillars.

This species is known in Berwickshire as "The Gowk;" and in some parts of the county the 1st of April is called " Hunt the Gowk,"[3] when boys and others are sent on fools' errands, and the following distich is sometimes heard—

> The first of April,
> Hunt the Gowk another mile.

The name of "Gowk oats" is given in Berwickshire to grain of that kind sown in April. In an early March it is said, "there will be no 'Gowk oats,'"[4] whilst in the late spring of 1888, when heavy snow-storms prevailed until the end of this month, farmers were heard remarking that "they would be a' 'Gowk oats' thegither this year." Before the adoption of the New Style in 1752, the Cuckoo would generally arrive in this county shortly after the middle of April, and this may have given rise to the association of the "Gowk" with that month.

---

[1] Mr. James Smail mentions that he saw a young Cuckoo at Burncastle, Lauderdale, in the end of August.—*Hist. Ber. Nat. Club*, vol. viii. p. 104.

[2] I have known a young Cuckoo shot at Broomhouse, near Duns, as late as the beginning of September.

[3] Jamieson says:—"Young people, attracted by the singular cry of the Cuckoo, being anxious to see it, are often very assiduous to obtain this gratification. But as the bird changes its place so secretly and suddenly, when they think they are just within reach of it they hear its cry at a considerable distance. Thus they run from place to place, still finding themselves as far removed from their object as ever. Hence the phrase 'Hunt the Gowk,' may have come to be used for any fruitless attempt, and particularly for those vain errands on which persons are sent on the first day of April."—*Scot. Dict.*, Art. "Gowk."

[4] "Popular History of the Cuckoo," by James Hardy.—*Folk Lore Record* vol. ii. p. 57.

In the Spottiswoode district, Mr. Lockie says, the following rhyme with reference to the Cuckoo is sometimes heard :—

> In April come he will,
> In May he sings all day,
> In June he changes tune,
> In July away he'll fly,
> In August go he must.

Mr. Hardy writes that it is popularly believed in Berwickshire, as elsewhere, that the Cuckoo keeps his voice in good tune only so long as he can get small birds' eggs to suck, and that when these are all hatched his musical notes grow hoarser and huskier until at length his melodious functions are entirely suspended.[1] He also says that "it is the common belief in the county that if the circumstances in which its note is first heard for the season be attended to, they afford unerring signs whereby the secrets of a man's destiny for the ensuing year may be disclosed. In whatever direction he may be looking when its tones arrest him, there he will be on the anniversary of that day next year. If he be gazing on the ground he is warned of an untimely fate. If he has money in his pocket, it is an omen that he shall not lack ; if penniless, that the cruse of oil shall not be replenished, and that losses and disappointments shall be his lot."[2] It is likewise a popular superstition that if, on hearing the first Cuckoo of the season, a girl takes off her shoe and examines it, she may find inside a hair of the colour of that of her future husband. Mr. Hardy gives an instance of a girl, who was carrying the seed-basket for the sower in one of his fields at Oldcambus on the 7th of April 1876, having been observed to do this.[3] The same practice appears to be

---

[1] " Popular History of the Cuckoo."—*Folk Lore Record*, vol. ii. pp. 59-60.
[2] *Ibid.* p. 90.
[3] *MS. Notes* by Mr. Hardy.

common in the west of Scotland and elsewhere. Gay thus refers to it in his "Shepherd's Week"—

> When first the year, I heard the Cuckoo sing,
> And call with welcome note the budding spring,
>
> . . . .
>
> Upon a rising bank I sat adown,
> And doff'd my shoe, and by my troth I swear,
> Therein I spied this yellow frizzled hair,
> As like to Lubberkin's in curl and hue,
> As if upon his comely pate it grew.

A story is told of some worthy inhabitants of the village of Gordon, who, wishing to have perennial spring and summer, thought to attain their object by building a high wall round a place frequented by the Cuckoo; but the bird escaped, and the "Gowks o' Gordon" consoled themselves by the reflection that the wall had not been built high enough.[1]

The ancient name of Godscroft, the estate of David Hume, the historian of the Douglases and the Homes, is said to have been "Gowkscroft," from the great number of Cuckoos which frequented it; and to this day the country people call the place "Gowkscraft," and sometimes "Gowkie." According to Dr. Henderson, of Chirnside, Quixwood means the wood of the Cuckoo, and appears to have been derived from the Celtic word, *Cuach*, the Cuckoo.[2]

Several wild plants which grow in Berwickshire are named from this bird. Dr. George Johnston, writing of *Cardamine pratensis*, says:—"Our children, with whom the plant is a favourite, call it also 'Cuckoo-Flower,' because the Cuckoo often drops what they believe to be its spittle on the leaves.[3] Wood Sorrel (*Oxalis acetosella*), a common plant in our woods, is called 'Cuckoo's-meat,' 'Gowk's-meat,' and 'Gowk's

---

[1] *Hist. Ber. Nat. Club*, vol. ix. p. 228. See also "Pop. Hist. of the Cuckoo," *Folk Lore Record*, vol. ii. pp. 67-68, by James Hardy.

[2] Dr. Henderson's *MS. Notes*.

[3] *Natural History of the Eastern Borders*, vol. i. "Botany," p. 33.

clover.'[1] *Luzula campestris*, which is common in pastures, and flowers with the primrose and dog-violet, and which is pulled by children to give variety to the spring nosegay, is called 'Cuckoo-grass;' while *Arum maculatum*, which is known as 'Cuckoo-pint,' is found plentifully in the woods about Blanerne and Netherbyres,"

> Where peep the gaping speckled cuckoo-flowers,
> Prizes to rambling school-boys' vacant hours.
> CLARE.

Some lingering remains of the ancient belief in the hibernation of the Cuckoo and the Swallow are occasionally heard amongst boys, who repeat the following lines :—

> Seven sleepers there be,—
> The Bat, the Bee, the Butterflee,
> The Cuckoo and the Swallow,
> The Kittiwake and the Corn-craik,
> Sleep a' in a little hollie.

---

[1] In Gothland, Sweden, it is called "Giokmat," and in France, "Pain de Coucou."—*Nat. Hist. Eastern Borders*, p. 50.

THE BARN OWL.

WHITE OWL, YELLOW OWL, SCREECH OWL, HISSING OWL,
CHURCH OWL, GILLIHOWTER, HOWLET, HOOLET.

Aluco flammeus.

The White Hoolet, The Hoolet.

The bird of eve flits sullen by,
Her home, these aisles and arches high ;
The choral hymn, that erst so clear
Broke softly sweet on fancy's ear,
Is drowned amid the mournful scream
That breaks the magic of my dream ;
Roused by the sound, I start and see
The ruin'd sad reality.
BURNS, *Lincluden Abbey.*

Solis et occasum servans de culmine summo
Nequidquam seros exercet Noctua cantus.
VIRGIL, *Georgics.*

THIS beautiful, interesting, and very useful bird, which was formerly found in various districts of Berwickshire, seems now to be almost extinct in the county. From twenty-five to fifty years ago it was plentiful, and used to breed regularly every year in the precipitous rocks on the banks of the Tweed, Whitadder, and Blackadder, as well as in some of our old ruins.

The high rocks on the north side of the Tweed at Scarsheugh, a short distance above Milne Graden, were a favourite resort, and here, Mr. Douglas, banker, Coldstream, informs me he remembers it nesting fifty years ago, when he was a boy. Miss Georgina Milne-Home, Milne Graden, speaks of its appearance at the same rocks at a much later date;[1] and, about twenty years ago, when a party of crow-shooters rowed up the Tweed to the Scarsheugh rocks, to shoot the wild dovecot Pigeons found there, the firing of the guns caused several "White Hoolets" to come out of the crevices, which are now only inhabited by Jackdaws, Starlings, and Pigeons.[2] Mr. J. Speedy states that the Barn Owl used to breed in a quarry by the side of the Tweed at Ladykirk, when he lived there, upwards of twenty years ago; and it is also said to have formerly reared its young in the rocks by the side of that river, near Birgham.[3]

About thirty years ago, this Owl nested annually in the "Pigeon Rock" on the north bank of the Whitadder at Edrington Castle; and young men from the adjoining village of Paxton were sometimes lowered over the cliff to get the Owlets when they were nearly fledged, for the purpose of keeping them as pets.[4] "Hoolet's Craig," on the Whitadder, near Cockburn, opposite the Staneshill, would appear to have been at one time a home of this bird, and there are other rocky precipices which would doubtless be frequented by it. Mr. W. Duns, builder, Duns, says that he recollects a pair of Barn Owls occupying a hole left for scaffolding in the wall of Duns Town Hall when it was built (about 1821), so common was the bird at that time.

A favourite haunt on the Blackadder, about thirty years ago, was the steep rock below the Conservatory at Blackadder House; and here the "White Hoolet" used to be seen

---

[1] In letter to me dated 9th January 1886.
[2] Verbal information from Mr. James Purves, gamekeeper, Mordington.
[3] Mr. George Bolam, Berwick-on-Tweed, *MS. Note.*
[4] Verbal information from Mr. Robert Dalgleish, Paxton.

coming from her nesting-place in the face of the precipice on summer evenings, to search, with silent wing, the adjoining grounds for mice and other prey, with which to feed her young.[1]

In the Lauderdale district, the drainage holes of Norton Bridge, near Thirlstane Castle, were occupied by this owl about forty years since; and it also bred at Old Norton,[2] in the neighbourhood of which it might then be seen skimming noiselessly over the haughs of the Leader and other streams, in the early twilight, and occasionally pouncing down upon some unfortunate mouse, which it would bear off in its claws with a scream of satisfaction.[3] The ruins of Corsbie Tower, in the parish of Legerwood—the ancient seat of the Cranstouns—were formerly a well-known resort of the White Owl, from which, in the dusk of the evening, she sallied out to Spottiswoode[4] and her other hunting-grounds in the vicinity, whence in due time she returned

In sullen silence to her ancient home.
WILSON, *Morning*.

Lady Grizel Baillie has informed me that, until 1841, a pair of Barn Owls had their nest every year in one of the chimneys of Mellerstain House; and Mr. Hardy, Oldcambus, writes that it used to be found in the Pease Dean.[5] It is

---

[1] Verbal information from Mr. James Renton, Allanbank, who has lived at Blackadder for upwards of forty years. Mr. Brunton, Clarabad Mill, tells me that he recollects a boy at Allanbank Stables, named Archibald Simpson, getting a young White Owl out of its nest in the rocks at Blackadder House, thirty years ago.

[2] *Hist. Ber. Nat. Club*, vol. vii. p. 302.

[3] Information from Mr. Peter Scott, Lauder.

[4] Information from Lady John Scott Spottiswoode, of Spottiswoode, who writes to Mr. P. Stormonth Darling:—"*Spottiswoode, 5th March* 1886.—I have been inquiring about the White Owls. They used to build every year in Corsbie Tower, and came about here from there. I remember when we were children my sister had a tame young White Owl, which was given to her by old Mr. Murray, the tenant of Corsbie."

[5] Mr. Hardy's *MS. Notes*.

said to have been formerly plentiful about The Hirsel, Kimmerghame, and Polwarth Kirk.[1]

From numerous inquiries made in every district of Berwickshire, the Barn Owl does not appear to have been known to breed anywhere in the county for upwards of twenty years. At the present time (April 1888) all its old nesting-places on the banks of the Tweed, Whitadder, and Blackadder, and in ruins, such as Corsbie Tower, are occupied by colonies of Jackdaws; so it is very probable that the great increase of these birds in Berwickshire within the last five-and-twenty or thirty years has been one of the causes of the almost complete extinction of this species. It is somewhat remarkable that while the Barn Owl has nearly disappeared, its congener, the Tawny Owl (*Strix aluco*), has seemingly suffered no diminution in numbers, if it has not increased, within the period mentioned; for it is a common bird in all our woods, where its loud hooting may be often heard during the night in autumn and winter. But the latter has not been subjected to the same struggle for existence as the former, for it lays its eggs in a hole in a tree, or in a squirrel's old nest in the woods, which are not liable to be invaded by colonies of Jackdaws, like the crevices in the rocky cliffs where the Barn Owl was wont to breed. It is also probable that the Arctic Winter of 1860-61[2] killed many of the Barn Owls then existing in the county, for they are much more susceptible to cold than the Tawny Owl.[3]

No Barn Owls have been seen for many years in some districts of Berwickshire where, formerly, they abounded.

---

[1] Information from Mr. Smith, gamekeeper, Duns Castle.

[2] During Christmas week of the winter of 1860-61 the snow lay thickly over the whole county, and the thermometer ranged from 2° to 7° below zero. For three weeks in succession the Tweed at Milne Graden was frozen over, the ice being from 7 to 10 inches thick.—*Hist. Ber. Nat. Club*, vol. v. pp. 234, 235.

[3] See Seebohm's *British Birds*, vol. i. p. 148.

Mr. Tillie, Lauder, informs me that he has not known of its occurrence in Lauderdale for the last thirty years. In the neighbourhood of Paxton it has not been observed since 1872, when a specimen was unfortunately shot on the banks of the Whitadder, near Clairvale. Mr. George Turnbull tells me that the last Barn Owl noticed at Abbey St. Bathans was found dead there about sixteen years ago. One was shot at Nisbet House in 1872, as it flew out of the dovecot; and in December 1878 another, which had perished in the snow, was discovered near the forester's house at the Grange Wood, in Coldingham Parish. The next record of the appearance of this bird in the county is of an example which was shot by Mr. W. Robeson, Langrig, as it rose out of a double hedge on the march between Ravelaw and Leetside, on the 27th of October 1886; and about that date a second was got near Chirnside Bridge. In the same year a Barn Owl was seen at Blanerne;[1] and on the 20th of December 1887, a beautiful specimen was observed in a plantation on Milne Graden estate, near the old Tollhouse at Skaithmuir.[2]

The Barn Owl inhabits rocky precipices on the banks of rivers, old ruins, towers, and barns, as well as hollow trees;[3] and it continues to occupy and breed in the same retreat for many years in succession, if not molested. Here it dozes from sunrise to sunset, when it leaves its place of repose to skim over the meadows and corn-fields or along hedgerows, in search of prey. It also visits stackyards and farm buildings; and, when it has young, it has been known to

---

[1] Information in letter from Mr. Bruce, gamekeeper, Blanerne, dated 27th April 1888.

[2] Information in letter from Mr. Alexander Miller, hedger, Milne Graden, dated 20th December 1887.

[3] Mr. David Richardson, gamekeeper, Broomhouse, informed me on the 2nd of June 1888, that about five-and-twenty years ago the White Owl bred regularly every year in a hollow tree there.

return to its nest every quarter of an hour with a mouse in its claws.

Its principal food consists of mice and rats, which it destroys in great numbers; and, as it thereby renders valuable services to the farmer, we find Grahame, in his *British Georgics*, saying:—

> Let the screeching Owl
> A sacred bird be held; protect her nest,
> Whether in neighbouring crag, within the reach
> Of venturous boy, it hang, or in the rent
> Of some old echoing tower, where her sad plaint
> The livelong night she moans, save when she skims,
> Prowling, along the ground, or, through your barn,
> Her nightly round performs; unwelcome guest!
> Whose meteor-eyes shoot horror through the dark,
> And numb the tiny revellers with dread.
> *September.*

The Barn Owl likewise feeds upon small birds and sometimes upon insects, the remains of which have been recognised on examination of the ejected pellets[1] which are found near its favourite haunts.

This owl does not hoot, its usual note being a loud scream, which, when heard at night in a lonely place, such as an ancient churchyard, or among the ruins of an old tower or abbey, adds to the weirdness of the scene, and recalls the graphic lines of Blair;—

> The wind is up: hark! how it howls! methinks
> Till now I never heard a sound so dreary:
> Doors creak, and windows clap, and night's foul bird,
> Rooked in the spire, screams loud: the gloomy aisles,
> Black-plastered, and hung round with shreds of 'scutcheons,
> And tattered coats of arms, send back the sound,
> Laden with heavier airs, from the low vaults,
> The mansions of the dead. Roused from their slumbers,

---

[1] Like other rapacious birds the White Owl ejects pellets, consisting of the indigestible parts of the animals and birds upon which it feeds. Mr. Seebohm mentions (*Hist. Brit. Birds*, vol. i. p. 149) that "out of seven hundred pellets of this Owl, which were carefully examined by Dr. Altum, remains were found of 16 bats, 2513 mice, 1 mole, and 22 birds, 19 of which were sparrows."

THE BARN OWL.

> In grim array, the grisly spectres rise,
> Grim, horrible, and obstinately sullen,
> Pass and repass, hushed as the foot of night.
> Again the screech-owl shrieks : ungracious sound ;
> I 'll hear no more ; it makes one's blood run chill.

The Barn Owl has also a hissing note, which it has been heard to utter when disturbed in its retreat during the daytime ; and the young are said to " snore " when they are hungry.

It is not such an early breeder as its congeners; the eggs, which are from three to seven in number, being usually found about the beginning of May. They are dull white in colour, and about the size of those of a common dovecot Pigeon, though not so much elongated.

This species may be easily distinguished from our other owls by its pure white breast, yellowish back, and black eyes.

THE LONG-EARED OWL.

HORNED OWL, HORNED HOOLET, HORNIE HOOLET.

Asio otus.

The Horned Hooler.

The hornyt byrd quhilk we clepe the nicht Oule,
Laithely of forme, with crukit camscho beik.
　　　　　　GAVIN DOUGLAS, *Description of Wynter.*

THE Long-eared Owl is a permanent resident[1] in most of the pine-woods throughout Berwickshire, and it also frequents small strips and clumps of firs by the roadsides. During the day-time, it usually sits concealed amongst the thickly-set branches towards the top of a tree, close to the trunk; and, as its mottled plumage somewhat resembles the colour of the bark, it is not easily noticed. Although it generally chooses a bushy fir for its retreat, it does not invariably do so, for, on the 29th of August 1874, I found two sitting in a large leafy hawthorn by the side of the Tweed, near Paxton House; and Mr. W. Duns, Duns, informs me that, a few years ago, he saw an Owl of this kind amongst the ivy which covers the old aisle in Ayton Churchyard.

This species does not appear to be much bewildered by sunlight, and, if disturbed while roosting in some thick tree

---

[1] The Long-eared Owl is occasionally seen on migration at the lighthouses on the east coasts of England and Scotland. It is possible that our Berwickshire birds may receive some additions to their numbers from migrants in autumn, and that these visitors may return northwards again in spring.

in the day-time, flies noiselessly away to another in the neighbourhood. In the dusk of the evening it leaves its place of concealment, to search for mice, rats, and small birds, upon which it chiefly preys.

Mr. Abel Chapman, who writes to me that he has attended to the habits of this Owl for many years, says:—"The only cry I ever heard it utter (except for a short period in summer), was a low cat-like whistle, almost a monotone, but occasionally varied with something very like the mee-owing of a cat. This was just before it left the woods for its nightly hunt at dusk. The other note I have referred to was heard only in summer, about June, when the young were newly fledged, and was a sort of petulant bark, somewhat like that of a petted lap-dog. It was not very loud, and not to be compared with the hooting of the Tawny Owl. I have several times seen the Long-eared Owl uttering this note, always when the bird was perched near the side of a wood, and I have thought it was uttered by way of encouragement or instruction to the young Owls."[1]

It generally selects the old nest of a Magpie,[2] Carrion Crow, Wood Pigeon, or Squirrel, in which to deposit its eggs, which are four or five in number, smooth and white, of the size of those of a tame Pigeon, but more oval in shape. The eggs are usually laid in the end of March, or in April.

---

[1] Mr. Abel Chapman, in letter, dated Roker, 29th March 1886.

[2] My friend, Mr. A. H. Evans, informs me that, in April 1876, he found the young of a Long-eared Owl in an old Magpie's nest in a small plantation by the side of the public road near Sunwick. They were five in number, and newly hatched. Some years ago it nested on a larch near the Lover's Tryste in the Crow Dean Wood at Paxton, a Wood Pigeon having her nest on the tree at the same time.

THE SHORT-EARED OWL.

WOODCOCK OWL, MARSH OWL, MOSS OWL.

Asio accipitrinus.

The Woodcock Owl.

> *The Owl by day,*
> *If he arise, is mocked and wondered at.*
> SHAKESPEARE, *Henry VI.*

THIS Owl is a winter visitor to Berwickshire, making its appearance with the Woodcock from the north of Europe in October and November, and leaving us for the northern regions with that bird in March; hence its popular name of Woodcock Owl.[1]

While on migration in autumn and spring, it is observed during night at the lighthouses on the east coasts of England and Scotland;[2] and, on arriving, it seems to settle in considerable numbers in certain localities near the sea-coast, as many as twenty having been found on Mordington Moor during a day's search for Woodcock in November.[3] It is frequently seen during this month in the neighbourhood of the Grange Wood, in the parish of Coldingham.

This species does not inhabit woods, rocky precipices, or old ruins, like our other Owls, but frequents moors and

---

[1] It is probable that when heavy snow-storms occur in the country in winter, this Owl emigrates further south, and visits us again in spring on its migration northwards.

[2] *Reports on the Migration of Birds,* 1879-86.

[3] Information from Mr. James Purves, gamekeeper, Mordington.

open fields, where it roosts on the ground. It is occasionally found by partridge-shooters during the autumn and winter months, amongst heather, ferns, or rushes, in turnips, or by the grassy side of a hedge or ditch. On being flushed during the day, it does not seem to be bewildered by the light, but generally flies off with a buoyant flight to some distance, and again alights amongst thick herbage on the ground. It very seldom perches on a tree.

The Short-Eared Owl feeds chiefly upon field-mice, for which it hunts moorlands, fields, and the sides of hedges, in the dusk of the evening; but it is sometimes seen quartering the ground in search of prey during the hours of daylight.

Some of the Border counties suffered greatly from immense swarms of field-mice[1] (*Arvicola agrestis*) from 1875 to 1877, the district most seriously affected being a cluster of farms at the head of Borthwick Water, which falls into the Teviot about three miles above Hawick. The mice destroyed the roots of the herbage[2] to such an extent that the sheep-farmers suffered great loss amongst their flocks, and various plans were tried to get rid of the enemy, with little success; but meanwhile many rapacious birds, including Short-Eared Owls, appeared on the scene, and helped to clear the little depredators off the ground. Sir Walter Elliot, in his interesting report on "The Plague of Field-Mice on the Border Farms in 1876-7," refers to the systematic destruction of all kinds of birds of prey (whose part in the economy of nature is to keep the smaller animals within due bounds) as having helped to produce the plague.[3]

---

[1] The excessive propagation of the field-mice at this period is believed to have been caused by the general mildness of the seasons from 1870-76.—*Hist. Ber. Nat. Club*, vol. viii. pp. 459, 460.

[2] This is called "Spret" by the border shepherds.—*Hist. Ber. Nat. Club*, vol. viii. p. 470.

[3] *Hist. Ber. Nat. Club.* vol. viii. pp. 447-468.

There can be no doubt that Owls do very great service to farmers, by killing immense numbers of mice as well as rats; and examinations of the pellets containing the indigestible parts of their prey have proved that they do very little damage to game. On this account it would certainly be for the interest of the landowners of the county to prevent the destruction of these useful birds on their estates.

As Mr. Robert Gray records that he saw a young owl of this species with traces of down upon it, which had been shot near Ayton on the 13th of July 1876,[1] it is probable that a few pairs of Short-Eared Owls remain in the county during summer, and breed on some of the unfrequented moors. It is said that one or two nests have been found in the heather on the Lauderdale hills in the month of June.[2] This Owl makes a scanty nest on the ground amongst heather, rushes, or other herbage, and the eggs, which vary from four to seven in number, are smooth, and pure white.

The Short-Eared Owl has been heard to utter a harsh scream when its nest is disturbed. It may be easily distinguished from our other Owls by the black feathers which surround its yellow eyes.

---

[1] *Hist. Ber. Nat. Club*, vol. viii. p. 156.
[2] *Ibid.* vol. vii. p. 302.

THE TAWNY OWL.

BROWN OWL, GREY OWL, WOOD OWL, IVY OWL, BEECH OWL, HOWLET, JENNY HOWLET.

Strix aluco.

The Hoolet, The Grey Hoolet, The Jenny Hoolet, The Roarer.

> The Houlet's eerie, midnight wail,
> Frae White-ha's[1] lone embowered dale,
> Portended yet some fearfu' tale
> Thou 'dst gar us hear,
> And make the staff o' life to fail
> I' the coming year!
>
>   DR. HENDERSON, *Winter Rhymes.*

THIS is the commonest Owl in Berwickshire, and its loud hooting is frequently heard resounding through the woods at night, in the autumn, winter, and early spring months. From April until about the end of August, it is generally silent; but in the dusk of summer evenings, the cry of the young Owlets after they leave the nest for the branches of some neighbouring tree, which resembles the words "Kee-wick, kee-wick," takes the place of the usual melancholy but pleasing note of their parents. With the advent of September, when the nights are becoming longer, the plantations again re-echo with the prolonged "Hoo-hoo-hoo" of the Tawny Owl. At this season, as the shades of evening fall over the valley of the Tweed at Paxton, and the harvest moon is seen rising majestically over the trees, a

---

[1] An estate on the banks of the Whitadder, near Chirnside, the property of Mr. Mitchell Innes of Ayton.

distant hoot may be heard in the recesses of the woods. This is quickly followed by another in the vicinity, and if the listener blow "mimic hootings" on his closed hands,[1] the whole of the Owls in the neighbourhood immediately join in chorus,

> with quivering peals
> And long halloos, and screams and echoes loud
> Redoubled and redoubled.
> WORDSWORTH.

Even during the day, "the tremulous sob of the complaining Owl" sometimes reaches the ear from its secluded retreat in the woods, where, perched high up amongst the branches of a Scotch or spruce fir, it sits dozing close to the trunk, until the evening.[2]

In former times the Owl was generally regarded as a bird of ill omen, and its "notes of woe" were heard with superstitious fear. Shakespeare makes its doleful cry pierce the ear of Lady Macbeth while the murder is being done, and calls it "the fatal bellman which gives the stern'st good night."

This species is much bewildered by sunlight; and the smaller birds, such as Blackbirds and Chaffinches, take a peculiar delight in mobbing and tormenting it, if they discover its hiding-place during the day. They fly round about, and dart into, the tree in which the object of their hatred is perched, screaming and chattering all the while; and the attack is generally continued until the Owl is com-

---

[1] On favourable nights, my friend Mr. Arthur H. Evans, of Cambridge, and I, have sometimes heard half-a-dozen Owls in different parts of the Paxton woods answering our calls, one after another, as quickly as they could hoot. An echo, which repeats the calls when they are made in my garden, adds to their effect. Sometimes all the Owls are silent, and will not answer a call. They hoot on certain nights, and are not heard on others.

[2] Miss Georgina Milne Home has informed me that, on the 6th of June 1886, she heard an Owl hooting near the Old Quarry, at Milne Graden, between one and two o'clock in the afternoon. I heard one hooting at Paxton, at 1.30 P.M., on the 14th of March 1886; also, at 2.30 P.M., on the 8th of February 1888.

pelled to change her retreat. Where this Owl is numerous, it may be frequently seen in the dusk of the evening flying silently round the edges of woods, over parks, or round stackyards, in search of its prey, which usually consists of mice, such as the long-tailed field-mouse, the short-tailed field-mouse, the common barn-yard mouse, shrews, moles, rats, beetles, and also small birds.

Like its congeners, it ejects by the mouth, in the form of elongated pellets, the skin, feathers, fur, and bones of its prey. These pellets are usually found in quantities about the roosting-places of Owls, and an examination of them shows the nature of their food.[1]

The benefit which this Owl confers upon farmers and landowners, by destroying enormous numbers of mice and other vermin, is very great; and, as it does little damage to game, it should be protected, and encouraged to increase.

It is an early breeder, and generally has eggs in March, or in the beginning of April. It usually selects a hole in a

---

[1] Professor Newton, in giving the following interesting and instructive table, remarks that "the infallibility of the evidence thus afforded as to the food of Owls is as complete as the way of obtaining it, by those who have the opportunity, is simple. Several German naturalists have made some very precise researches on this subject. The following results, with regard to our three commonest species of Owls, are those afforded by the investigations of Dr. Altum, as communicated by him to the German Ornithologists' Society during its meeting in 1862:—

	No. of Pellets Examined.	Remains Found.						
		Bats.	Rats.	Mice.	Voles.	Shrews.	Moles.	Birds.	Beetles.
Tawny Owl, .	210	..	6	42	206	33	48	18 *	48
Long-Eared Owl,	25	..	..	6	35	..	..	2 †	..
Barn Owl, ..	706	16	3	237	693	1590	..	22 §	..

* 1 Tree-Creeper, 1 Yellow Bunting, 1 Wagtail, 15 small species undetermined.
† Species of Titmouse.   § 19 Sparrows, 1 Greenfinch, 2 Swifts.
Besides a countless number of Cockchaffers.—Yarrell, *History of British Birds*, fourth ed., vol. i. pp. 147, 148.

THE TAWNY OWL.

tree for its nest;[1] but sometimes lays in the deserted habitation of a Magpie or Crow on the branches of a Scotch fir, or in a crevice of a rocky precipice. The eggs, which are from three to five in number, are considerably larger and more oval than those of the Wood Pigeon, smooth, and pure white.

For some time after they are hatched, the young Owlets are covered with a greyish white down, and, when first able to fly, they sit about the trees in the vicinity of the nest, where the old birds continue to feed them, and where in summer evenings they utter their cry of "Kee-wick, kee-wick."

Houlet Ha', a farm on Spottiswoode estate, within a mile of Westruther village, seems to have derived its name from this bird.

---

[1] It used to breed in a hole in a large beech-tree at Finchy, on the Tweed, near Paxton; and Mr. Compton-Lundie of Spital tells me that a favourite nesting place at Spital House is the interior of an old hollow ash. Mr. W. Duns, Duns, says he has known the nest in a hole in a plane-tree at Nisbet, every season for many years past. It has been seen coming out of holes in the rocky precipice of Scarsheugh, on the Tweed, near Milne Graden, during the breeding season.

THE HEN-HARRIER.

COMMON HARRIER, BLUE GLED, BROWN GLED, WHITE-ABOON GLED, BLUE KITE, BLUE HAWK, SEA-GULL HAWK.

Circus cyaneus.

𝔗𝔥𝔢 𝔊𝔩𝔢𝔡,[1] 𝔗𝔥𝔢 𝔊𝔯𝔢𝔶 𝔊𝔩𝔢𝔡.

The mittane[2] and Sanct Martin's Fowl,[3]
Wend[4] he had been the Hornit Owl,
They set upon him with a yowl,
And gave him dint for dint.

DUNBAR, *Of the Fenzeit Friar of Tungland.*

IN olden times, when all the lower parts of the county were covered with marshes, bogs, and stagnant pools, as described by Mr. John Wilson, late of Edington Mains, in a Report on the Agriculture of Scotland,[5] the Hen-Harrier doubtless bred in the Merse in considerable numbers, for it would find suitable hunting-grounds among the numerous swamps, where it would also make its nest and rear its young in security. During the great advance of agriculture which took place in Berwickshire from the middle to the end of last century, most of the morasses were

---

[1] From the Anglo-Saxon *glidan*, to glide.—Jamieson, *Scot. Dict.*

[2] Mittane, a bird of prey of the hawk kind.—Jamieson, *Scot. Dict.*

[3] M. Eugène Rolland says, in his *Faune populaire de la France* (tom. ii., "Les Oiseaux sauvages," p. 24), with regard to the Hen-Harrier being called in that country L'Oiseau de Saint Martin :—" Il est appelé ainsi parce qu'il effectue son passage à travers la France vers le 11 Novembre, jour de la Saint-Martin."

[4] "Thought."

[5] "Report on the Present State of the Agriculture of Scotland, arranged under the auspices of the Highland and Agricultural Society, to be presented at the International Congress at Paris in June 1878."

drained, with the exception of Billie Mire, which, down to the beginning of the present century, remained in its natural state. This vast bog, of which a full account is given in the article on the Bittern, lay along the greater part of a narrow valley which extended from near Ayton, on the east, to the vicinity of Chirnside, on the west—a distance of about five miles. During summer its surface was covered with great beds of luxuriant reeds, bulrushes, and other rank water-plants; while here and there might be seen small clumps of lichen-covered grey-saughs (*Salix cinerea*), and alders, flourishing on the drier parts of the ground. The deep black pools, which were to be found towards the centre of the quagmire, and which were regarded by the inhabitants of the neighbouring hamlets with feelings of superstitious awe, as the haunt of "Jock o' the Mire," abounded with frogs, newts, and leeches. Innumerable waterfowl, and other birds of various kinds, found a congenial home here, and in the midst of the reeds and rushes, the Hen-Harrier nested every year until about 1830-35, when the Mire was at last effectually drained.

Mr. White, farmer, Lennelhill, informs me that he lived for many years in the early part of this century, near the eastern extremity of Billie Mire, having been taken there in 1815, when he was three years old, by his father, who had then entered upon a lease of the farm of Causewaybank. Although some attempts had been previously made to drain the Mire, it remained much in its original state during his boyhood, and he remembers that the Gled used to be seen flying about the bog in summer, and that he and the boys who herded the cows, which pastured on the drier parts of the marsh, used to find its nest every season amongst the sedges and rushes in the swamps. After the final drainage of the Mire, which, he says, took place about

1830-35, the Gleds forsook the neighbourhood. Mr. William Allan, farmer, Bowshiel, who, in his youth, lived at Billie Mains with his father,[1] also stated to me that this bird bred during his boyhood in the western part of Billie Mire, where it was frequently seen skimming over the rushes in search of prey. He recollects that the nest, which he sometimes found amongst the bog-reeds of the marsh, was piled up with sticks, the foundation being made with thorns, and that he had to wade through water half-way up his legs to get at it. Mr. Thomas Hewit, Auchencrow, has mentioned to me that his brother, who lived at Auchencrow Mains, found a nest with young in Billie Mire, on the east side of the road leading from Auchencrow to Causewaybank, and that one of the fledglings was taken home and kept until it was full grown, when it attacked some of Mr. Logan's[2] chickens, and had to be destroyed. He relates that the boys about Auchencrow Mains, in his youth, used to visit the Mire on Sundays, to search for the Gled's nest; and, on these occasions, they provided themselves with long "leading-in" ropes from the farm, which they tied round their waists before venturing on the treacherous surface of the bog, giving the loose end into the hands of some of their companions, who remained on the firm ground at the side, ready to pull them out if necessary.

Although Billie Mire seems to have been the principal haunt of the Hen-Harrier in Berwickshire down to about 1830, its eggs were then occasionally found in the uplands of the county also, whilst on the Lammermuirs it appears to have bred in considerable numbers, until a much later date. Mr. Cowe, Oldcastles, states that about 1820,

---

[1] Mr. Allan was born at Billie Mains about 1806, and lived there until 1827, when he went to Bowshiel.

[2] Mr. Logan was tenant of Auchencrow Mains Farm.

when he was a boy, the Gled used to build its nest on the ground in a field called the Whinny Park,[1] on Wester Muirside Farm, then occupied by his father; and I find Mr. Wilson, Coldingham, noting in 1865, that it had been known to breed at Outlawhill, and was said to have done so at one time about Earnsheugh.[2] Mr. Robert Renton, Spottiswoode Lodge, who, in 1835, lived at Jordanlaw, says he remembers a Gled's nest being found about 1837 amongst rushes at a clump of birks to the north of Jordanlaw farmsteading; and Mr. Hardy, when on a visit to Lauderdale in the beginning of July 1842, records that Lord Lauderdale's forester informed him that this species was sometimes seen on the moors there, where it nested amongst the heather.[3] At that date, Mr. Peter Scott, Lauder, tells me the Hen-Harrier was a well-known bird on the surrounding hills, and that he found a nest with five young on the flat moor at the top of Wheelburn, about half-way between Lauder and Broadshawrig. The female was shot near the nest, for when any person approached she hovered round it, but the male was shy and kept out of range. When his father came from Whittinghame, in East-Lothian, as gamekeeper to Lord Lauderdale, about 1837, Hen-Harriers were numerous on the Lauderdale moors, and sometimes four or five specimens might be seen nailed up in a row on the vermin-rail near his father's house. They were honoured with three nails—one through the head, and one at each tip of the outstretched wings. Mr. Kelly mentions that this bird was once common on the Lammermoors, where its nest was usually built with heather "birns."[4] It would appear from the reports of the Penmanshiel

---

[1] Mr. Cowe says this park is now a part of Bogaugreen Farm.
[2] Mr. Hardy's *MS. Notes*.
[3] *Ibid.*
[4] *Hist. Ber. Nat. Club*, vol. vii. p. 302.

shepherds, to have frequented the moors there long ago.[1] Mr. Abraham Mack, Abbey St. Bathans, has told me that when he was a boy, about 1826, he used to see big Grey Gleds in the dean between Blackburn and Bowshiel, where one was shot and given to Mr. W. Aitchison, Cockburnspath. Mr. William Patterson, late of North Berwick Abbey Farm, has informed me that he killed a Hen-Harrier at Greenhead, in the parish of Coldingham, in the winter of 1845, while snow lay on the ground. It had caught a partridge when he first observed it. A few years afterwards, his brother obtained a beautiful male on the farm of Blackhill, in the same parish. Mr. Hardy records that in his boyhood, about 1820-30, he often saw the Grey Gled on Coldingham Moor.[2]

Dr. Henderson, Coldstream, has informed me that, in 1872, a female Hen-Harrier was shot in the woods behind Rumbleton Law, by Mr. Robert Henderson, East Gordon. On the 16th of October 1874, a bird of the same species and sex, was seen by me in a plantation at the side of the Tweed, near Paxton. It rose from the branch of a tree on which it had been sitting apparently asleep, for it allowed me to approach within thirty yards of its perch, and on rising it hovered away over towards the Whitadder, at a considerable height in the air. A female was killed near the Hunter's Well, on Quixwood Moor in January 1877, and is in possession of Mr. John Hogg of Quixwood, who

---

[1] Mr. Hardy says in his *MS. Notes*:—"The Gled must have been numerous formerly in the Cockburnspath district. The ancient name of Ewieside, according to the tradition of the very old people, was Gledstane Forest. The Gled's Stane was a large stone near a road leading across the hill, that stood on the hill top on the boundary between Cocklaw and the Tower, and was removed, to the disapproval of many old people, by the tenant of Bowshiel, a century ago, probably to build some cottars' houses at Ecklaw, which were then being erected. Sir John Hall regarded the few old trees on the east end of Ewieside as the remains of an old forest, and forbade their being cut down. These notes were taken in 1855, from information by an old shepherd who was brought up at Bowshiel." He adds that a game at which children play is called "Shoogled-wylie."

[2] *Hist. Ber. Nat. Club*, vol. vii. p. 247.

had it preserved. Another example, of the same sex, was shot at Drakemire on the 22nd of December 1888.

The Hen-Harrier is generally seen skimming over the surface of the ground on buoyant wing in search of prey, which consists of small quadrupeds, birds, and reptiles. It is very destructive to game, especially young Grouse and Partridges, and, on this account, it has been exterminated in Berwickshire.

It nests on the ground, usually in the heather on a moor, or amongst whins or other low bushes on waste land; and when a reed-bed in a marsh is selected for its breeding quarters, the nest is piled up with sticks and sedges to a little height above the surface of the bog, to protect the eggs and young from the water with which the swamp is liable to be flooded after heavy rains. The eggs, which are four or five in number, are usually bluish white, without spots, but they are sometimes slightly marked with yellowish brown.

It is probable that the Kite (*Milvus ictinus*) was found in Berwickshire in former times, and it would, along with the Hen-Harrier and the Buzzard, be known as the Gled. In addition to Gledstane Forest, previously mentioned, the Gled appears to have given its name to Gladswood, on the Tweed above Dryburgh, and Gladscleugh Burn which joins Whalplaw Burn near Peat Law (1367 ft.), in the parish of Lauder.

There is no record of any specimen of the Marsh Harrier (*Circus æruginosus*), or Montagu's Harrier (*Circus cineraceus*), having been obtained in Berwickshire.

THE BUZZARD.

COMMON BUZZARD, GLEAD, GLED, KITE, PUTTOCK.

Buteo vulgaris.

> *The sheep-boy whistled loud, and lo!*
> *That instant, startled by the shock,*
> *The Buzzard mounted from the rock,*
> *Deliberate and slow.*
> WORDSWORTH.

ALTHOUGH it is probable that, in former times, the Buzzard bred in Berwickshire, it is now only an occasional autumn and winter visitor to the county, when on migration from the northern parts of Europe.

Selby, writing in 1839, mentions an example which he had received from Mellerstain.[1] Lauderdale appears to be a favourite resort of this bird, for we find Mr. Kelly recording that six or seven specimens had been obtained there, from about 1868 to 1875,[2] and that two occurred in the spring and winter of 1876, while one was seen at Blythe Edge in January 1877.[3] Mr. Hardy relates that a Buzzard was observed about the beginning of October 1874, moving with slow, leisurely flight, between the Pease Bridge and Cockburnspath,[4] and that another, in immature plumage, was captured about two miles north-east of Duns, on the 22nd of February 1877.[5] Mr. Charles Watson, Duns,

---

[1] *Hist. Ber. Nat. Club*, vol. i. p. 256.
[2] *Ibid.* vol. vii. p. 301.
[3] *Ibid.* vol. viii. p. 142.
[4] *Ibid.* vol. vii. p. 294.
[5] *Ibid.* vol. viii. p. 196.

has informed me that one was shot at Nisbet, in May 1878; and in October 1872, Mr. Turnbull, of Abbey St. Bathans, showed me a beautiful specimen, which had been killed a few years before that date by Mr. W. Mark Elliot, on the march between Primrosehill and The Retreat. Mr. Hunter, of Anton's Hill, wrote to me on the 18th of January 1886, that a Buzzard had been shot two days previously by his gamekeeper, on the banks of the Leet, below Belchester House. An unusual number of birds of this species visited the county in the following autumn. One was observed for some days in the third week of October, frequenting the neighbourhood of Oatleycleugh and Cockburn Law; and about the 22nd of that month it was caught by the gamekeeper at Cockburn. Having seen it rise from a rabbit in a trap, which it had killed on the moor, by the side of the Eller Burn, he set the trap close to the prey, and the Buzzard was caught in the course of an hour, on returning to its repast. On the 12th of November following, another was taken in a similar manner on Kimmerghame estate, in a field adjoining the Berwick road; and, on the afternoon of the same day, while driving to Wedderburn, I saw one alight on a tree by the side of the public road leading from Nabdean to Fishwick. It was followed by two Rooks which were buffeting it, and, after resting for a short time on its perch, within thirty yards of my horse's head, it flew slowly away to the Paxton woods. A few weeks afterwards, a bird of this kind was seen in the Pistol Plantations by Dr. Stewart.

The Buzzard feeds upon rabbits, moles, mice, and other small quadrupeds, as well as birds and reptiles. It seems to have a capacious throat, for, in a specimen dissected by Mr. Walter Simson, Lauder, a mole was found, which had been swallowed entire.[1] It is not very destructive to

---

[1] *Hist. Ber. Nat. Club*, vol. vii. p. 301.

winged game, preying chiefly upon any which may be wounded or weak; but nevertheless, we find it included in an Act by James II. of Scotland in 1457—" anent Ruikes, Crawes, and uther foules of riefe, as Eirnes, Bissettes, Gleddes, Mittalles, the quhilk destroyis baith cornes and wilde foules, sik at Pertrickes, Plovares, and utheris."[1]

It varies considerably in the colour of its plumage, some specimens having much more white upon them than others; but it may be generally distinguished when flying, by the upper parts being brown, with much white under the wings.

---

[1] *Laws and Acts of Parliament made by the Kings and Queens of Scotland*, by Sir Thomas Murray of Glendook: Edinburgh, 1681.

THE ROUGH-LEGGED BUZZARD.

ROUGH-LEGGED FALCON.

Buteo lagopus.

The monarch bird, with blytheness hard
The chaunting litil silvan bard,
Calit up a Buzart, quha was then
His favorite and chamberlane.

RAMSAY, *Eagle and Robin.*

THE Rough-Legged Buzzard sometimes visits Berwickshire in the autumn and winter months, when it is on migration from the north of Europe.

Writing in 1875, Mr. Kelly says:—"Twenty years ago Mr. Simson, Lauder, added this bird to his collection. It was caught, he tells me, by the shepherd of Huntington, who, while engaged one day looking over the fields, came suddenly upon it eating a Rabbit. At his approach the bird moved off slowly and, as he thought, reluctantly, from the half-finished meal to a neighbouring wood. This hesitation encouraged him to set a trap and wait for the result, which he had not long to do, for, in a short time, the marauder returned and was caught."[1] Ten years later a second specimen was captured near Lauder in a similar way by a rabbit-catcher; and, on the 13th of February 1877, another was trapped by Lord Lauderdale's gamekeeper in a field near Edgarshope.[2]

---

[1] *Hist. Ber. Nat. Club*, vol. vii. p. 301, 302.
[2] *Ibid.* vol. vii. p. 302.

Mr. Robert Gray, the well-known author of *The Birds of the West of Scotland*, records a male shot near Coldingham about the end of October 1875,[1] and a female killed in the same locality on the 1st of March 1877.[2] A Rough-Legged Buzzard frequented Lamberton Moor for about a fortnight in December 1878, being sometimes seen in the Lang Belt Plantation there; and, after the heavy snowstorm came on, it used to visit the stackyard at Lamberton Shiels, apparently for the purpose of catching mice.[3] A specimen was shot while pursuing a grouse on Quixwood Moor on the 15th of January 1880,[4] and another was killed by the gamekeeper at Ladykirk, near Bendibus, on the Tweed, in December of the same year.[5]

When this species visits Berwickshire it seems to prefer the opener districts of the county, and especially those abounding with rabbits, upon which, with smaller quadrupeds, such as mice,[6] and likewise birds and reptiles, it chiefly preys.

The Rough-Legged Buzzard, which varies considerably in its plumage, may be easily distinguished by its legs being feathered down to the toes.

---

[1] *Hist. Ber. Nat. Club*, vol. vii. p. 463.
[2] *Ibid.* vol. viii. p. 155.   [3] *Ibid.* vol. viii. p. 503.
[4] Information from Mr. Hogg of Quixwood.
[5] Information from Mr. Basset, gamekeeper, Ladykirk.
[6] During the plague of mice on some of the pastoral farms in Roxburghshire, in 1876-77, as many as six or seven Rough-Legged Buzzards were sometimes seen on the wing at once, in places where the mice were most plentiful.—*Hist. Ber. Nat. Club*, vol. viii. p. 456.

THE SEA EAGLE.

WHITE-TAILED EAGLE, GREY EAGLE, CINEROUS EAGLE, ERNE.[1]

Haliæetus albicilla.

> *Like to an Eagle in his kingly pride,*
> *Soaring through his wide empire of the aire*
> *To weather his brode sails.*
> SPENSER, *Faerie Queen.*

THE Sea Eagle, which is the largest of our rapacious birds, seldom visits the county.

Selby, in his Report on the Ornithology of Berwickshire, says:—"Several instances of the Great Sea Eagle have occurred within our precincts. All the examples of this kind that I have seen and examined have been in immature plumage; a circumstance, however, not at all remarkable, as the adults, when once paired, rarely leave the immediate vicinity of the eyry they have selected, and the young, after quitting the nest, are always driven from the district in which they have been bred by the parent birds."[2]

The poet Thomson refers to this characteristic habit of the Eagle in the following beautiful lines:—

> High from the summit of a craggy cliff,
> Hung o'er the deep; such as amazing frowns
> On utmost Kilda's shore, whose lonely race
> Resign the setting sun to Indian worlds,
> The Royal Eagle draws his vigorous young,

---

[1] Ern, Erne, Elrne, Earn :--the Eagle. Anglo-Saxon, Earn.—Jamieson's *Scot. Dict.*

[2] Report on the Ornithology of Berwickshire, and the district within the limits of the Berwickshire Naturalists' Club.—*Hist. Ber. Nat. Club*, vol. i. p. 250.

> Strong pounced, and ardent with paternal fire,
> Now fit to raise a kingdom of their own,
> He drives them from his fort, the towering-seat
> For ages of his empire, which, in peace,
> Unstained he holds; while many a league to sea
> He wings his course, and preys in distant isles.

The Rev. Andrew Baird, who, in December 1834, wrote the Report on the united parishes of Cockburnspath and Oldcambus, in the *New Statistical Account of Scotland*, mentions that the "Sea Eagle has been occasionally shot;"[1] and Dr. Robert D. Thomson, in his account of the parish of Eccles, in the same year, says that it sometimes visits that parish.[2] An Eagle frequented Dabb's Head, a hill in Lauderdale, for some time in the spring of 1844.[3] A specimen was trapped on South Fallaknowe, in the parish of Coldingham, in March 1866.[4] Writing in the last-mentioned year, Mr. W. P. Turnbull remarks that this bird has been frequently observed at St. Abb's Head;[5] and Mr. Robert Gray states, in 1871, that it is occasionally seen in autumn about the precipitous cliffs there.[6] In the autumn of 1872 an Eagle, apparently of this species, was observed on the top of Huntlaw (1625 feet), on the boundary between the parishes of Lauder and Longformacus; and in December of that year a specimen was trapped at Charter-

---

[1] *New Statistical Account of Scotland*, vol. ii. (Berwickshire), p. 291.

[2] *Ibid.* vol. ii. (Berwickshire), p. 53.

[3] Information from Mr. Johnston, farmer, Huntington, Lauder. Dabb's Head is one of the most conspicuous points of the Lammermoor Hills, on the old farm of Earnscleugh, which is now incorporated with the farm of Burncastle, near Lauder. It is, according to the Ordnance Survey Map of Berwickshire, 1257 feet above the level of the sea.

[4] This specimen was erroneously reported in the local newspapers of the time as a Golden Eagle, and Mr. Turnbull, in his *Birds of East-Lothian* (p. 9), records it as such. Mr. W. Paterson, late of North Berwick Abbey Farm, who is well acquainted with British birds, of which he had a large collection, wrote to me on the 8th of May 1886, that he saw the example referred to, at the house of the gamekeeper who trapped it, shortly after it was caught, and found it to be a fine specimen of the White-Tailed Eagle.

[5] *Birds of East-Lothian*, 1866, p. 10.

[6] *Birds of the West of Scotland*, 1871, p. 13.

hall, in the parish of Fogo, where it was kept in confinement until the 30th of September 1886, when it died.[1] On the 7th of February 1877, a Sea Eagle in immature plumage was shot near Bedshiel, in the parish of Greenlaw, of which Mr. W. Smith, gamekeeper, Duns Castle, has given me the following account. He informs me that a short time previous to the above date he observed a large bird which he supposed to be a Heron rise out of a ditch which leads from the Hule Moss to Marchmont, but on looking at the spot where the bird rose he found a hare lying on the ground half-eaten, and it then occurred to him that the bird which he had seen was an Eagle. About a week afterwards, when he and his brother were out shooting rabbits on Bedshiel and Hallyburton, they noticed a large bird at some height in the air, approaching them from the north, and as they conjectured that it was the Eagle, and would be sure to alight upon the "Sappers and Miners'" pole, which was not far from the place where they stood, they lay down on the heather, when, as anticipated, the bird came flying in the direction of the pole, and hovered over it. They both had double-barrelled muzzle-loading guns, and as the weather was wet, the caps on three of the barrels missed fire, but the fourth, which was fired by his brother, brought down the bird.[2] He states that three other Eagles were seen about the Dirringtons that season.

The food of this species consists chiefly of carrion of every kind, stranded fish, hares, rabbits, and winged game, as well as the young of sea-fowl.

The subject of our notice can be easily distinguished

---

[1] Information from Mr. William Thompson, gamekeeper, Charterhall, on the 11th of October 1886.
[2] The specimen here referred to was kindly presented by Sir Hugh Hume-Campbell of Marchmont, Bart., to the Berwick Museum, where it may now be seen.

from the Golden Eagle by the tarsi of the former being bare, whilst in the latter they are covered with feathers down to the toes.

It is probable that this bird was much more frequently seen in Berwickshire in ancient times than now, and may have given its name to Earnsheugh—a precipice on the sea-coast about a mile to the west of St. Abb's Head —Earnscleugh Rig Hill (1122 feet), and Earnscleugh Water, in the Lammermoors, near Lauder. Mr. Hardy says that the old name of Piperton Hill, near Oldcambus, was Earnslaw, which is also the designation of a farm about a mile south-west of Swinton. A small homestead near Whitsome is called Eaglehall.

There is no authentic record of the occurrence of the Golden Eagle (*Aquila chrysaëtus*) in Berwickshire.

THE SPARROW-HAWK.

BLUE HAWK, PIGEON HAWK, GLEG HAWK.

Accipiter nisus.

𝕮𝖍𝖊 𝕾𝖕𝖆𝖗𝖗𝖔𝖜-𝕳𝖆𝖜𝖐.

Enough for me
To boast the gentle Spar-Hawk on my fist,
Or fly the Partridge from the bristly field,
Retrieve the covey with my busy train,
Or with my soaring Hobby dare the Lark.

SOMERVILLE, *Field Sports.*

IN most of the wooded districts of the county the Sparrow-Hawk is frequently seen, and more especially in summer, when it is rearing its young, for it is then so bold that it often ventures into our gardens and farm-steadings in search of prey. At this season the gamekeeper has to be on the alert, if he be rearing young pheasants, for the Sparrow-Hawk will seize and carry them off within a few yards of the watcher, and, if not prevented, will return again and again, at short intervals, to steal the chicks. Mr. Thomas Speedy, who was for some years assistant gamekeeper at Ladykirk, says:—"We have known cases where Pheasants were being hand-reared, and where this little impudent thief succeeded in carrying off twenty or thirty young birds before it was shot. With the rapidity of lightning, it dashed in from some adjacent rock or wood, and although the keeper was on the watch with gun in hand, it succeeded in clutching its unsuspecting prey before the mother-bird could give an indication of warning, and was out of range before an aim could be secured."[1]

The flight of this bird is very rapid, and when it is seen gliding along the side of a hedge or the edge of a wood, within a few feet of the ground, the wings appear to be only half-expanded. It is capable of turning very quickly on the wing, and may be sometimes observed darting into hedges and bushes to seize small birds which have flown there for shelter from their enemy.

> Unequal now the chace! struggling she strives,
> Entangled in the thorny labyrinth,
> While easily its way the small bird winds,
> Regaining soon the centre of the grove.
>
> GRAHAME, *Birds of Scotland.*

When sweeping along near the ground in search of prey, it has the habit of occasionally curving upwards in its flight, to alight upon a tree or wall, apparently for the purpose of resting, or for observation, and on these occasions it sits almost upright upon its perch. When it visits farm-buildings and stackyards, its presence is generally indicated by the alarm of all the small birds in the neighbourhood, such as Sparrows, Chaffinches, and Greenfinches, which hasten to conceal themselves in the nearest hedge or thicket, where, after the danger is past, they make loud outcries, the Sparrows chirping and the Chaffinches "twinking" incessantly for some time. The dove-cot Pigeons, too, dread its approach, and fly round and round the steading in great alarm. The prey of the Sparrow-Hawk chiefly consists of small birds, such as Yellowhammers, Greenfinches, and Chaffinches; but the female Hawk, which is considerably

---

[1] *Sport in the Highlands and Lowlands of Scotland*, by Thomas Speedy, 1884, p. 366.

larger and more powerful than the male,[1] will take Partridges,[2] young Pheasants, Pigeons, leverets, and young rabbits. One day in November 1885, when shooting with Mr. Clapham on the banks of the Whitadder at Broomhouse, near Duns, I observed many places in the plantations and by the hedge sides, where Wood Pigeons had been killed and plucked by this Hawk, the feathers being scattered in great profusion round the spot where its prey had been devoured.

This species appears to have been frequently used in Falconry in olden times, to take the smaller kinds of game, and we find it occasionally mentioned in the Accounts of the Lord High Treasurer of Scotland with reference to this sport. Thus, on the 16th of September 1473, the following entry occurs—

"Item gevin to a man of Dauid Oguiluiys of Inchmartyne that brocht a Spar Halk to the King,[3] . . . . iij s

Dame Juliana Berners, in her "Boke of St. Albans," mentions that according to the laws of Falconry different kinds of hawks were assigned to different ranks, the priest being entitled to carry the Sparrow-Hawk (the female),[3] while to the holy-water clerk was assigned the Musket.[4]

"As a bird of the chase," says Mr. Belany, "the Sparrow-Hawk has maintained a considerable reputation. For a

---

[1] Rolland, in his *Faune populaire de la France*, p. 36, says :—"On dit d'un individu qui épouse une femme plus forte, plus intelligente, plus riche qui lui : 'Il fait un mariage d'épervier, la femelle vaut mieux que le mâle.'"

[2] On the 28th of October 1887, I saw a female Sparrow-Hawk rise from a Partridge which she had killed, and partly devoured, at the side of a hedge, on the farm of Ancrum Woodhead, near Jedburgh.

[3] *Accounts of the Lord High Treasurer of Scotland*, vol. i. 1473-98, p. 45.

[4] The name given by falconers of old to the male Sparrow-Hawk. It is mentioned by Shakespeare in the *Merry Wives of Windsor*, Mrs. Ford addressing Falstaff's page with "How now, my Eyas-Musket?" An Eyas-Musket is a young male Sparrow-Hawk taken from the nest.
Rolland, in his *Faune populaire de la France*, referring to the Sparrow-Hawk, says "Il a en commun avec d'autres oiseaux de proie les noms suivants :—

Mosquet, *M.* ancien Provençal. Muxet, *M.* Catalan des Pyrénées-Orientales.

And he adds, " En fauconnerie, le mot *mouchet* s'applique à mâle seulement."

short distance its flight is rapid in the extreme, and it will take Partridges, Magpies, Landrails, Water Hens, and other smaller quarry, in a manner by no means inferior to some of the Falcons. It gets sooner upon the wing, or darts off the first with more rapidity than the Falcons, and is therefore better suited for bush-hawking." [1]

The nest, which is constructed of sticks and slender twigs like that of a Wood Pigeon, but larger, is generally found in a plantation, and usually in a Scotch, or spruce fir. Sometimes, however, the eggs, four or five in number, bluish white, blotched with reddish brown, are deposited in the deserted nest of a Magpie or Crow. The old Heronry Wood at Paxton is a favourite resort of this Hawk, and here its eyry has been frequently found in some of the tall Scotch firs, one being discovered, with four young, in June 1886.

---

[1] *Falconry*, by J. C. Belany, 1841, p. 169.

John Blair 1887. The Pech Stane

THE HONEY BUZZARD.

BROWN BEE-HAWK, CAPPED BUZZARD.

Pernis apivorus.

> Some haggard Hawk who had her eyry nigh,
> Well pounced to fasten, and well wing'd to fly,
> One they might trust their common wrongs to wreak,
> The Musquet and the Coystrel were too weak ;
> Too fierce the Falcon ; but above the rest,
> The noble Buzzard ever pleased me best ;
> Of small renown 'tis true, for, not to lie,
> We call him but a Hawk by courtesy.
>
> — DRYDEN, *Hind and Panther.*

THE Honey Buzzard is a rare visitor to Berwickshire. Mr. Robert Gray mentions that three or four specimens were shot in this county in June 1845, where the species again reappeared in one or two instances in 1863-64 ;[1] and, according to Mr. Andrew Brotherston, one was obtained at Newton Don on the 22nd of May 1865.[2] Mr. Hardy writes that—" A very fine example was killed on the 29th of May 1876, in Penmanshiel Wood, by a gamekeeper on Dunglass estate. His attention was called to it by the persecution which it met with from the carrion crows, of which about twenty were attacking it while it was perched on the top of a tree, whence it sallied forth on its sable tormentors, which then fled with wild outcries. It flew sluggishly, like an owl. I examined it shortly after it was shot, and it is now pre-

---

[1] *Birds of the West of Scotland*, p. 49.
[2] *Hist. Ber. Nat. Club*, vol. vii. p. 135.

served by the gamekeeper at Dunglass."[1] Its stomach was filled with wasps, honey-bees, bumbees (humble-bees), and beetles.[2] In the summer of 1879 a bird of this species frequented Penmanshiel Wood, and was left unmolested.[3] A beautiful specimen, with cream-coloured under-parts, was shot in the south plantation at Laws, in the parish of Whitsome, by Mr. John Alder, on the 26th of October 1888.

The flight of this Buzzard is slow, like that of its congeners, and it resembles them in its general habits, but it differs from them considerably with regard to its food, which principally consists of the larvæ of wasps, bees, and other insects, as well as earth-worms, slugs, small birds' eggs, and moles.

The plumage varies much in colour, especially in birds of the first and second year, and, according to Mr. Seebohm, there appear to be light and dark forms. He adds that this species may be at once distinguished from the buzzard (*Buteo vulgaris*) "by the scales on the tarsus, which are finely reticulated all round, instead of being in broad plates at the front and the back. Another equally important distinction may be found in the lores, which are finely feathered down to the cere, in place of being only covered with bristles."[4]

---

[1] *Hist. Ber. Nat. Club*, vol. viii. pp. 190, 191.
[2] Information from Mr. John Bolton, gamekeeper, Dunglass, in letter dated the 3rd of December 1881.
[3] *Hist. Ber. Nat. Club*, vol. ix. p. 409.
[4] Henry Seebohm, *History of British Birds*, vol. i. (1883), p. 71.

I.—INDEX TO NAMES OF BIRDS IN VOL. I.

LATIN NAMES.

	PAGE
Acrocephalus schœnobænus,	75
,, nævius,	79
Accentor modularis,	83
Acredula caudata,	89
Anthus pratensis,	113
,, trivialis,	116
,, obscurus,	118
Ampelidæ,	129
Ampelis garrulus,	129
Alaudidæ,	250
Alauda arvensis,	250
Alcedinidæ,	267
Alcedo ispida,	267
Aluco flammeus,	280
Asio otus,	287
Asio accipitrinus,	289
Accipitres,	296
Accipiter nisus,	311
Buteo vulgaris,	302
Buteo lagopus,	305
Cinclidæ,	85
Cinclus aquaticus,	85
Chelidon urbica,	141
Cotile riparia,	144
Certhiidæ,	146
Certhia familiaris,	146
Carduelis elegans,	148
,, spinus,	152
Coccothraustes chloris,	154
Corvidæ,	197
Corvus monedula,	207
,, corone,	210
,, cornix,	214
,, frugilegus,	216
,, corax,	240

	PAGE
Cypselidæ,	256
Cypselus apus,	256
Caprimulgidæ,	258
Caprimulgus europœus,	258
Cuculidæ,	273
Cuculus canorus,	273
Circus cyaneus,	296
Dendrocopus major,	261
Erithaca rubecula,	50
Emberizidæ,	180
Emberiza miliaria,	180
,, citrinella,	183
,, schœniclus,	186
Fringillidæ,	148
Fringilla cœlebs,	162
,, montifringilla,	166
Falconidæ,	296
Garrulus glandularius,	200
Hirundinidæ,	135
Hirundo rustica,	135
Haliæetus albicilla.	307
Iynx torquilla,	265
Laniidæ,	123
Lanius excubitor,	123
,, collurio,	127
Linota cannabina,	168
,, rufescens,	172
,, flavirostris,	174
Loxia curvirostra,	179

INDEX TO NAMES OF BIRDS: LATIN.

	PAGE		PAGE
Motacillidæ,	108	*Sylviidæ,*	36
Motacilla lugubris,	108	*Saxicola œnanthe,*	36
,, *sulphurea,*	111	*Saxicola rubetra,*	41
Muscicapidæ,	131	*Saxicola rubicola,*	44
Muscicapa grisola,	131	*Sylvia rufa,*	57
,, *atricapilla,*	133	,, *atricapilla,*	60
		,, *salicaria,*	63
Oriolidæ,	120	*Sittidæ,*	100
Oriolus galbula,	120	*Sitta cæsia,*	100
		Sturnidæ,	192
Passeres,	1	*Sturnus vulgaris,*	192
Phœnicura ruticilla,	47	*Striges,*	280
Phylloscopus collybita,	68	*Strigidæ,*	280
,, *trochilus,*	71	*Strix aluco,*	292
,, *sibilatrix,*	73		
Paridæ,	89		
Parus major,	91	*Turdidæ,*	1
,, *ater,*	93	*Turdus musicus,*	1
,, *palustris,*	95	,, *viscivorus,*	8
,, *cæruleus,*	97	,, *iliacus,*	14
Passer domesticus,	156	,, *pilaris,*	18
,, *montanus,*	160	,, *varius,*	22
Pyrrhula europœa,	176	,, *merula,*	25
,, *enucleator,*	178	,, *torquatus,*	32
Plectrophanes nivalis,	189	*Troglodytidæ,*	102
Pyrrhocorax graculus,	197	*Troglodytes parvulus,*	102
Pica rustica,	202		
Picidæ,	261		
		Upupidæ,	271
Regulus cristatus,	65	*Upupa epops,*	271

ENGLISH NAMES.

Blackbird,	25	Cole Titmouse,	93
Blackcap,	60	Creeper, Tree,	146
Blue Titmouse,	97	Chaffinch,	162
Brambling,	166	Crossbill,	179
Bullfinch,	176	Corn Bunting,	180
Bunting, Corn,	180	Chough,	197
,, Yellow,	183	Carrion Crow,	210
,, Reed,	186	Crow, Carrion,	210
,, Snow,	189	,, Hooded,	214
Barn Owl,	280	Cuckoo,	273
Buzzard,	302		
,, Rough-Legged,	305	Dipper,	85
,, Honey,	315		
Chiffchaff,	68	Eagle, Sea,	307

INDEX TO NAMES OF BIRDS: ENGLISH.

	PAGE		PAGE
Fieldfare,	18	Pied Wagtail,	108
Flycatcher, Spotted,	131	Pipit, Meadow,	113
,, Pied,	133	,, Tree,	116
		,, Rock,	118
Garden Warbler,	63	Pied Flycatcher,	133
Golden-Crested Wren,	65	Pine Grosbeak,	178
Grasshopper Warbler,	79		
Great Titmouse,	91	Redwing,	14
Grey Wagtail,	111	Ring Ouzel,	32
Golden Oriole,	120	Redstart,	47
Great Grey Shrike,	123	Redbreast,	50
Grey Shrike, Great,	123	Rock Pipit,	118
Goldfinch,	148	Red-Backed Shrike,	127
Greenfinch,	154	Redpoll, Lesser,	172
Grosbeak, Pine,	178	Reed-Bunting,	186
Great Spotted Woodpecker,	261	Rook,	216
		Raven,	240
		Rough-Legged Buzzard,	305
Hedge Sparrow,	83		
House Sparrow,	156	Song Thrush,	1
Hooded Crow,	214	Stonechat,	44
Hoopoe,	271	Sedge Warbler,	75
Hen-Harrier,	296	Sparrow, Hedge,	83
Harrier, Hen,	296	Shrike, Great Grey,	123
		,, Red-Backed,	127
Jay,	200	Spotted Flycatcher,	131
Jackdaw,	207	Swallow,	135
		Sand Martin,	144
Kingfisher	267	Siskin,	152
		Sparrow, House,	156
Long-Tailed Titmouse,	89	,, Tree,	160
Linnet,	168	Snow-Bunting,	189
Lesser Redpoll,	172	Starling,	192
Long-Eared Owl,	287	Skylark,	250
		Swift,	256
Missel Thrush,	8	Short-Eared Owl,	289
Marsh Titmouse,	95	Sea Eagle,	307
Meadow Pipit,	113	Sparrow-Hawk,	311
Martin,	141		
,, Sand,	144	Thrush, Song,	1
Magpie,	202	,, Missel,	8
		,, White's,	22
Nuthatch,	100	Titmouse, Long-Tailed,	89
Nightjar,	258	,, Great,	91
		,, Cole,	93
Ouzel, Ring,	32	,, Marsh,	95
Oriole, Golden,	120	,, Blue,	97
Owl, Barn,	280	Tree Pipit,	116
,, Long-Eared,	287	Tree Creeper,	146
,, Short-Eared,	289		
,, Tawny,	293		

	PAGE		PAGE
Tree Sparrow,	160	Willow Wren, . . .	71
Twite, . .	174	Wood Wren, . . .	73
Tawny Owl, .	292	Warbler, Sedge, . . .	75
		Warbler, Grasshopper, .	79
White's Thrush, . .	22	Wren, ...	102
Wheatear, . . .	36	Wagtail, Pied, . . .	108
Whinchat, . . .	41	,, Grey, . . .	111
Whitethroat, . .	57	Waxwing, . . .	129
Warbler, Garden, .	63	Woodpecker, Great Spotted, .	261
Wren, Golden-Crested, .	65	Wryneck, ...	265
,, Willow, . .	71		
,, Wood, . .	73	Yellow Bunting, .	183

BERWICKSHIRE NAMES.

Big Mavis,	8	Hempie, .	. 83
Blackie,	25	Hemp Sparrow,	. 83
Bee-Eater,	91	Heather Lintie,	. 174
Blackhead,	91	Hoodie, .	210, 214
Bitterbank, .	144	Huddie Craw,	. 214
Bark-Speeler,	146	Hoolet, .	280, 292
Bunting, .	180	,, Horned,	. 287
		,, Grey,	. 292
Cock o' the North,	166	,, Jenny,	. 292
Corn Buntling, .	180	Horned Hoolet,	. 287
Coal Hood, . .	186	Hawk, Sparrow,	311
,, Hooden, .	186		
Corbie Craw, .	210	Jenny-Cut-throat,	57
Craw, . . .	216	Jay Piet, . .	200
Corn Craw, .	216	Jack, . .	207
Corbie, . .	240	Jenny Hoolet,	292
Cran, . .	256		
		Kaittie, . .	102
		Kaittie-Wa-Wren.	102
Feltie, .	. 8, 18	Kay, . . .	207
Feltifleer,	. 8, 18		
		Lady-Linty-White.	57
Grey Cheeper,	113, 118	Lintie, Green, .	154
Goldie, . .	148	,, Grey, .	168
Gooldie, .	148	,, Rose, .	168
Gooldspink, .	148	,, Whin, .	168
Green Lintie, .	154	,, Heather, .	174
Grey Lintie, .	168	Laverock, .	250
Gowk, . .	273		
Grey Hoolet,	292	Mavie, . .	1
Gled, . .	296	Mavis, . .	1
,, Grey, .	296	,, Big, .	8
Grey Gled, .	296	Miller's Thumb, .	65

INDEX TO NAMES OF BIRDS: BERWICKSHIRE.

	PAGE		PAGE
Nightingale, Scotch,	75	Titlin,	113
Night Hawk,	258		
Ox-e'e,	91, 97	Upper Shealer,	162
Piet,	202	White-Rump,	36
Pyet,	202	Whinchacker,	41
Pyot,	202	Whaty-whey-beard,	57
		Whuskie,	57
Rock Starling,	32	Whuskie-whey-beard,	57
Robin,	50	Weary,	65
Rose Lintie,	168	Wheary,	65
Red-Nebbed Crow,	197	Willy-mufl,	71
Roarer,	292	Waiter Craw,	85
Starling, Rock,	32	Water Wagtail,	108
Stanechacker,	36, 44	Willy Water Wagtail,	108
Scotch Nightingale,	75	Woodlark,	116
Seed Bird,	108	Window Swallow,	141
Sea Swallow,	141	Whin Lintie,	168
Sparry,	156	White Hoolet,	280
Shilfa,	162		
Shulfie,	162	Yellow Yite,	183
Snawfleck,	189	,, Yorlin,	183
Stirlin,	192	,, Hammer,	183
Sparrow-Hawk,	311	,, Yirlin,	183

II.—INDEX TO PROVERBS,

POPULAR RHYMES, SAYINGS AND WEATHER LORE ABOUT BIRDS.

CARRION CROW: superstitions, 212, 213.
CHAFFINCH: weather prognostic, 163.
CUCKOO: Gowk oats, 276; Hunt the Gowk, 276; rhymes and superstitions, 277, 278, 279; Gowks o' Gordon, 278.
GLED: name of children's game, 300.
LINNET: proverb, 169.
MAGPIE: rhyme, 203: superstitions, 203, 204; derivation of "Tale Piet," 206.
REDBREAST: weather prognostics, 52, 53; superstitions, 53.
RAVEN: superstitions, rhymes and proverbs, 240, 241, 242, 243.
ROOK: weather-lore, 235, 237; sayings and rhymes, 236, 238.

SKYLARK: sayings and weather-lore, 251; maxime of Archibald, sixth Earl of Angus while at Billie Castle, 251.
SNOW BUNTING: weather prognostics, 190.
STONECHAT: rhyme, 46.
SWALLOW: weather prognostic, 139; superstitions, 139, 140; saying, 140.
TAWNY OWL: formerly regarded as a bird of ill-omen, 293.
WHITETHROAT: belief amongst boys, 58.
WREN: belief, 105, 106; rhymes, 106, 107.
YELLOW BUNTING: superstitions, 184, 185.

III.—TOPOGRAPHICAL INDEX TO VOL. I.

Abbey St. Bathans, occurrence of Sand Grouse, xx.; Dotterel in former times, xxii.; Ring Ousel on moors during grouse-shooting season, 33; Whinchat at Allerburn, 42; Stonechat at Lady's Pocket Wood, 45; Redstart in woods, 48; Dipper's nest, 87; Reed-Bunting caught, 187; Magpie not allowed to be extirpated, 203; Parish Rook Report, 221; favourite resort of the Nightjar, 258; Nightjar shot, 260; Great Spotted Woodpecker seen, 263; haunt of the Cuckoo, 274; Barn Owl found dead, 284.

Aikieside Wood, Tree Pipit, 116.

Ale Water, Grey Wagtail, 111.

Allanbank, Chirnside, Barn Owl nesting, 282.

Allanbank, Lauder, Great Grey Shrike, 124; rookery, 231.

Allanton, Lesser Redpoll, 173.

Allanton Bridge, Grey Wagtail, 112; Sand Martin, 144.

Allanton Free Church, White Chaffinch, 164.

Allerburn, Whinchat, 42; Buzzard trapped, 303.

Allergrain Wood, Hooded Crow nesting, 214.

Auchencrow, Corn Buntings in flock, 181; Mr. Thomas Hewit's account of Gled in Billie Mire, 298.

Auchencrow Mains, Great Grey Shrike, 124.

Ayton, Billie Mire, xix.; Parish Rook Report, 221; Corbie's nest at Corbie Heugh, 242; young Short-Eared Owl shot in summer, 291; extent of Billie Mire, 297.

Ayton Castle, Blackbirds roosting in young fir woods, 27; Starlings roosting in evergreens, 193; Great Spotted Woodpecker shot, 263; Kingfisher on the Eye, 270.

Ayton Churchyard, Long-eared Owl in ivy, 287.

Ayton Cocklaw, rookery, 221; Nightjar shot, 260.

Barony of Bonkle, Archibald, 6th Earl of Angus, and his maxime about the Skylark and freedom, 251.

Barrowmill Woods, Rookery, 222; Nightjar common in former times, 259.

Bartlehill, Rookery, 226.

Bedshiel, Grouse formerly very scarce, xxii.; rookery in former times, 229; Sea Eagle killed, 309.

Belchester, Rookery, 225; Buzzard shot, 303.

Bell Hill, Cockburnspath, Ring Ouzel's arrival in spring, 32.

Bemerside, xiv.; the loch a favourite haunt of wildfowl in olden times, xvii.

Bendibus, Rough-legged Buzzard shot, 306.

Berrybank, Rookery, 223.

Billie Brae, Reed-Bunting, 187.

Billie Burn, Kingfisher, 269.

Billie Castle, Rookery here in former times, 221; Skylark, and maxime of Archibald, 6th Earl of Angus, about freedom, 251.

Billie Mains, Blackbird singing at 2.15 a.m., 30.

Billie Mire, a resort of wildfowl in olden times, xvii.; former haunt of Bittern, xix.; Sedge Warbler, 76; Reed-Bunting, 187; Hen-Harrier nesting in former times, 297, 298.

TOPOGRAPHICAL INDEX. 323

Birchy Bank, Nightjar's nest, 259.
Birgham, xiv.; Wren nesting in skeleton of Carrion Crow, 104; Barn Owl breeding in rocks at side of Tweed, 281.
Biter Cove, a former haunt of the Chough, 198.
Blackadder, lowest temperature in Scotland, 20; Redstart plentiful, 48; Long-Tailed Titmouse in Pistol Plantation, 89.
Blackadder House, cliffs on the Blackadder, now a resort of Jackdaws, 207; and formerly a haunt of the Barn Owl, 281, 282; Rookery, 226.
Blackadder Water, Barn Owl nesting in precipices, xx.; sources of, xxi.; resort of Willow Wren, 71; Water Ouzel's nest, 87; Grey Wagtail's haunt, 111; Sand Martins' nests, 145; Siskin on banks, 153; Corbie Heugh Quarry derives name from Raven, 242; Raven nesting in precipices, 248; Kingfisher at Greenlaw, 270; Barn Owls' old nesting-places, 283.
Blackburn, Grey Gled in dean, 300.
Blackburnrigg, a favourite haunt of Nightjar, 259.
Blackburnrig Wood, Wood Wren, 74.
Blackcastle Rings, Sand Martins' nests, 145.
Blackerston, Great Spotted Woodpecker killed, 263.
Blackhill, Hen-Harrier captured, 300.
Black Hill, one of the heights of the Lammermuirs, xxi.
Blackhill, Earlston, Ring Ouzel, 34; Raven trapped, 246.
Blackpotts, Raven pecking out sheep's eyes, 246.
Blainslie, Rookery, 231.
Blanerne, Rookery, 221, 222; Barn Owl seen, 284.
Blanerne Bridge, Spotted Flycatcher, 131.
Bleakheugh, gathering place of Martins before autumn migration, 143.
Blue Braes, Raven's nest, 245.

Blyth, Lauderdale, Ring Ouzel frequently seen, 34.
Blyth Water, Grey Wagtail, 111.
Blythe Edge, first record of Redstart in Berwickshire, 48; Buzzard seen, 302.
Bogangreen, Rookery, 224.
Boondreigh Burn, visited by Kingfisher, 269.
Borthwick, "Peepers" caught in horsehair "girns," 174.
Bowshiel, a resort of the Snow-Bunting, 190.
Bowshiel Dean, Missel Thrush's nest, 11; Grey Gled shot, 300.
Braid Bog, near Penmanshiel, a favourite haunt of Wild Geese, xxiii.
Brander Cove, a resort of the Green Cormorant, xxv.
Brunta Burn, a tributary called Pyatshaw Burn derives name from Magpie, 204.
Brockholl, "venyson" reserved in lease of fifteenth century, xv.
Broadshawrig, Ring Ouzel's resort, 34; Hen Harrier's nest, 299.
Broomdykes, Chiffchaff, 69.
Broomhouse, Chiffchaff, 69; Pied Wagtail nesting in bomb-shell, 110; Grey Wagtail's nest amongst sweet violets, 112; Swallowdean, 143; Magpie formerly numerous, 202; young Cuckoo shot in September, 276; Barn Owl breeding in hollow tree, 284; Wood Pigeons killed by Sparrow-Hawk, 313.
Buchan's Moor, formerly a resort of the Nightjar, 259.
Buncle, Parish Rook Report, 221, 222.
Buncle Woods, Missel Thrush found there in 1845, 8.
Burncastle, young Cuckoo seen in end of August, 276.
Burnhouses, Whinchat, 42; Chiffchaff, 69; Wood Wren, 74; Bullfinch, 176.
Burnmouth, Martin frequenting rocks on coast, 142; Raven's nest in former times, 245; flocks of Skylarks, 253; Kingfisher seen, 270.

Butterdean, breeding quarters of Martin on the Eye, 145; Goldfinches numerous long ago at Butterdean Mill, 148.
Buxley Dean, much frequented by Goldfinches in former times, 149.
Byrecleugh, large bag of Grouse, xxii.; Ring Ouzel known as Rock Starling, 32; Pyatshaw Ridge, 204; weather lore about Rook, 237; Raven shot, 247.
Byrecleugh Ridge, xxi.
Byrewalls, Red-Backed Shrike seen, 127.

Cabby Burn, Kingfisher's nest, 268.
Caldra, near Fogo, Sand Martin, 145.
Campfield, Lamberton Moor, Hoopoe seen, 272.
Captain's Plantation, Eccles, Rookery, 226.
Carfrae Mill, large bag of Grouse, xxii.
Carolside, nest of Long-tailed Titmouse, 90; haunts of Jackdaw, 208.
Castlelaw, Rookery, 224.
Cattleshiel, Wheatear, 37.
Cauldshiel, Lauder, Rookery, 231.
Causewaybank, Hen Harrier at Billie Mire, 297.
Channelkirk, xiii.; Parish Rook Report, 222.
Channobank, Ring Ouzel on whin bushes, 33; Great Spotted Woodpecker shot, 262.
Chapel, xiv.; nest of Long-Tailed Titmouse, 90; haunt of Jackdaw, 208; rookery 231.
Cheviot Mountains, xv.
Chesters, near Fogo, Rookery, 228.
Chirnside, xiv.; Billie Mire, xix.; Missel Thrush driving away Song Thrush, 13; Blackbird's spring notes, 30; old adage about Pyet's nest, 206; Parish Rook Report, 222; extent of Billie Mire in vicinity, 297.
Chirnside Bridge, Chiffchaff, 69; Wood Wren, 74; Barn Owl captured, 284.
Clairvale, Barn Owl shot, 284.
Clarabad Mill, Dipper's nest, 87; Yellow Wagtail (*M.raii*) seen, 112.

Cleekhimin, Kingfisher, 269.
Clockmill, haunt of Siskin, 153.
Cockburn, Nightjar captured in September, 260; "Hoolett's Craig," 281.
Cockburn Mill, Kingfisher's nest, 269.
Cockburn Law, Fieldfares feeding on rowans, 19; arrival of Whinchat in spring, 41; Stonechat, 45; Goldfinches in neighbourhood, 149; Buzzard trapped, 303.
Cockburnspath, xiii.; Lammermuirs, xxiii.; arrival of Whinchat in spring, 41; Great Titmouse called the "Bee-eater," 92; Pied Wagtail in spring, 108; Meadow Pipit arriving on sea-coast, 114; Tree Pipit frequents vicinity, 116; occurrence of Waxwing, 129; Corn Bunting seen, 180; Parish Rook Report, 223, 224; Raven building in sea cliffs, 244; Nightjar, 259; Hoopoe, 272; Buzzard seen, 302; Sea Eagle occasionally shot, 308.
Cockburnspath Cove, Stonechat, 45.
Cockburnspath Tower, resort of Swift, 256.
Cockburnspath Tower Dean, haunt of Sedge Warbler, 77; breeding quarters of Sand Martin, 145; resort of Siskin, 153.
Cockit Hat Plantation, on Lamberton Moor, Ring Ouzel observed, 34.
Coldingham, Magpie frequently seen, 203; Parish Rook Report, 223, 224; Ravens observed at St. Abb's Head, 246; Barn Owl at Grange Wood, 284, and Short-Eared Owl there, 289; Sea Eagle trapped at South Fallaknowe, 308.
Coldingham Church, Martins assembling on roof previous to autumn migration, 143.
Coldingham Coast, migration of Golden-Crested Wren, 66.
Coldingham Loch, Stonechat on hills, 45; Sedge Warbler, 77.
Coldingham Moor, a favourite resort of the Dotterel in former times, xxii.; Sedge Warbler haunting mosses, 77; resort of Twite, 175, and of the Reed-Bunting, 187; Ravens shot, 247; Grey Gled seen, 300.

Coldingham Sands, Redwing, 16.
Coldlands, flock of Corn-Buntings, 181.
Coldstream, xiv.; effects of severe winter of 1878-9 on Song Thrush, 5; Bohemian Waxwing caught, 129; Parish Rook Report, 224; Kingfisher, 267.
Corbie Hall, Fogo, 242.
Corbie Heugh, popular rhyme about, 241; Raven's nest at, 242.
Corbie Heugh Quarry, 242.
Corbie Hill, 242.
Cornysyke, popular rhyme about, 241.
Corsbie Boy, Whinchat, 42.
Corsbie Muir, Snow-Bunting, 190.
Corsbie Tower, Lesser Redpoll breeds in neighbourhood, 173; Starling nests in ruins, 195; Jackdaw plentiful, 207; Rookery, 231; a former home of the Barn Owl, 282, 283.
Coveyheugh House, Kingfisher's nest, 270.
Cowdenknowes, xiv.; Long-tailed Titmouse feeding at window, 89; its nest, 90; Tree Pipit, 117; Magpie, 203; Nightjar, 260.
Crawburn, 239.
Crawcleugh, 239.
Crawchill, South, 239.
Crawlaw, near Wedderlie, 239.
Crawlea Plantation, Foulden, 238.
Crawlee Burn, Gordon, 239.
Crawlee, Greenlaw, 238.
Cranshaws, name of Rawburn derived from Roe Deer, xv.; Ring Ouzel numerous, 34; Parish Rook Report, 225; names of places in parish derived from Rook, 239; Raven Craig and the Brownie o' Cranshaws, 243.
Craw's Entry, 239.
Crib Law, xxi.
Crossrigg Farm, Whinchat, 42.
Crow Butts, Blackpotts, 239.
Crowbutts, Chirnside, 239.
Crow Dean Wood, Paxton, Twite seen, 174; Long-Eared Owl's nest, 288.
Crow Green, 239.
Crowshiel, 239.
Crow Wood, St. Leonards, 238.

Crunkley, a resort of Goldfinches long ago, 149.
Cuddy's Stele, nest of the Nightjar, 259.
Cumledge, Rookery, 225; Nightjar, 259.

Dabb's Head, Lauderdale, Sea Eagle seen, 308.
Damford Shiel, Kingfisher breeding, 267.
Denewood, "venyson" reserved in a lease of the fifteenth century, xv.
Dirrington Law, Great, Ring Ouzel on "Glitters," 33.
Dirringtons, three Eagles observed, 309.
Dod Mill, Pyatshaw Wood, 204; Kingfisher on Boondreigh Burn, 269.
Dog Bush, Twite, 174.
Dowlaw, a favourite resort of the Dotterel, xxiii.; arrival of Ring Ouzel in spring, 32; and of Whinchat, 41, 42; Sedge Warbler in willow thickets, 77; Crossbill shot, 179; resort of Snow-Bunting, 190; "Rooks" and "Little Rooks" off coast, 239; Raven's nest, 244.
Dowlaw Burn, Grey Wagtail, 111.
Dowlaw Dean, Ring Ouzel's arrival in spring, 32; lines to memory of Alexander Allan Carr, surgeon in Ayton, by Dr. Henderson, Chirnside, 33; Stonechat nests, 45; Sedge Warbler noted in 1843, 77; Tree Pipit, 116; favourite haunt of Nightjar, 259.
Drakemire, cream-coloured Skylark observed, 254; Hen-Harrier shot, 301.
Dreeper Island, Milne Graden, cream-coloured Blackbird seen, 28.
Dryburgh Abbey, xiv.; Tree Pipit, 117; Starling nests, 195; Jackdaw breeds, 207; a favourite resort of the Swift, 256.
Drygrange, xiv.
Dunglass Burn, xiii.
Dun Law, one of the heights of the Lammermuirs, xxi.

TOPOGRAPHICAL INDEX.

Duns, effects of winter 1878-79 on Song Thrush, 5; cream-coloured Blackbird, 28; Nuthatch killed, 100; Pied Flycatcher shot, 133; superstition about Magpie in former times, 203; Parish Rook Report, 225; popular rhyme concerning Corbies, 241; Hoopoe killed by Cat, 271; Barn Owl nesting in Town Hall, 281; Buzzard captured near, 302.

Duns Castle, Redstart common, 48; Chiffchaff, 69; Waxwing shot, 130; Siskin, 153; Lesser Redpoll, 173; Jay killed, 201.

Duns Castle Lake, Sedge Warbler's haunt, 77.

Dye Cotttage, Ring Ouzel feeding on rowans, 34; Wheatear on road from Wedderlie, 37; Sand Martin's nest, 145; Jay killed, 201.

Dye Water, sources of, xxi.; Ring Ouzel found near upper streams, 33; Redstart in birch woods, 48; a favourite resort of Willow Wren, 71; haunt of Water Ouzel, 86; its nest, 87; Grey Wagtail, 111; Sand Martins, 145; flock of Twites, 175; Jackdaw's Craig, 208; Kingfisher, 270.

Dyksgatehead, popular belief about "Hoodie," 213.

Eaglehall, 310.

Earlston, Brambling, 167; Siskin, 173; Bullfinch, 176; Reed-Bunting, 187; Snow-Bunting, 190; Parish Rook Report, 225.

Earlston Blackhill, Ring Ouzel, 34.

Earnsheugh, near St. Abb's Head, the highest cliff on eastern seaboard of Great Britain, xxiv., xxv.; Raven's nest, 246; Hen-Harrier breeding, 299; name derived from Sea Eagle, 310.

Earnscleugh Glen, Ring Ouzel's nest, 35; Sea Eagle, 308.

Earnscleugh Rig Hill, name apparently derived from Sea Eagle, 310.

Earnscleugh Water, Sea Eagle, 310.

Earnslaw, 310.

Earnslaw, Piperton Hill, arrival of Whinchat, 41; haunt of Stonechat, 45.

East Blanerne, Sand Martin, 145; Brambling, 167; Mr. Blackadder's Report on rookeries in neighbourhood, 221, 222; Kingfisher on Billie Burn, 269.

Easter Harelaw, Walter Chisholm, 45.

East-Lothian, xiii.

East Mains, Lauder, flock of Goldfinches, 149, 150.

Eccles, Missel Thrush very rare in former times, 8; White's Thrush killed in parish, 22; Pied Blackbird, 29; Great Grey Shrike, 123, 124; occurrence of Pine Grosbeak, 178; Parish Rook Report, 225, 226; Sea Eagle, 308.

Ecklaw, Cuckoo, 274.

Eden Water, sources of, xxi.

Edgarhope Wood, Rookery, 230; Raven trapped, 247; Great Spotted Woodpecker killed, 262; Meadow Pipit feeding young Cuckoo, 275; Rough-Legged Buzzard caught, 305.

Edington Mains, Missel Thrush scarce long ago, 8; Great Grey Shrike, 123; flocks of Snow-Buntings, 190; Starling very rare in beginning of this century, 192; Raven's nest at "Blue Braes," 245; Mr. Wilson's description of Merse, 296.

Edingtonhill Wood, Great Spotted Woodpecker shot, 262.

Edington Mill, Siskin, 153; Lesser Redpoll, 173.

Edmonsdean, Whinchat breeds, 42; also Pied Wagtail, 109.

Edrington Mill, Martin breeds on steep rock, 142; banks of Whitadder frequented by Goldfinches, 150; Raven's Knowe, 242.

Edrington Castle, Jackdaw, 207; Barn Owl, 281.

Edrom, remains of Red Deer found in parish, xv.; Wood Wren, 74; Parish Rook Report, 226, 227.

Eildon Hills, xv.

Elbaw, Kingfisher's nest, 269.

Ellemford, Whinchat, 43; Stonechat, 45.

Ercildoun, xiv.

TOPOGRAPHICAL INDEX.   327

Evelaw Tower, colony of House Sparrows, 159; Starling nesting, 195; a favourite resort of the Swift, 256.

Ewieside, Ring Ouzel breeding, 33; haunt of Stonechat, 45; and of Nightjar, 259; a favourite resort of Cuckoo, 274.

Eyemouth, Redwings and Fieldfares, 15; Parish Rook Report, 227; Great Spotted Woodpecker shot, 262.

Eyemouth Fort, Hoopoe shot, 271.

Eye Water, sources of, xxi.; Whinchat, 42; Chiffchaff spreading in valley, 69; Willow Wren, 71; Wren, 103; Pied Wagtail, 109; Grey Wagtail, 111; Sand Martin, 145; Kingfisher's nest, 269, 270.

Fairneyside, Snow Bunting, 190; Raven's nest in rocks in olden times, 245.

Fallsidehill, Rookery, 229.

Fangrist Burn, Corbie Heugh Quarry, 242.

Fans, effects of severe winter of 1878-79 on Song Thrush, 5; Great Grey Shrike found dead, 124; Siskin, 153.

Farne Islands, Migration Reports; Song Thrush, 4; Blackbird, 27; Wheatear, 37; Redbreast, 51; Golden-Crested Wren, 66; Long-Tailed Titmouse, 89; Blue Titmouse, 98; Wren, 102; Pied Wagtail, 109; Meadow Pipit, 115; House Sparrow, 152; Chaffinch, 163; Lesser Redpoll, 172; Snow-Bunting, 189, 190; Starling, 195; Carrion Crow, 210; Hooded Crow, 214; Skylark, 252.

Fast Castle, xiii., xxiv.; Ring Ouzel in Dowlaw Dean, 33; Wren amongst rocks, 103; Martins' nests, 142; Chough formerly bred in cliffs, 197, 198; and now extinct, 199; a former resort of Raven, 243; Raven's nest, 244.

Flot Carr, xxv.

Finchy Shiel, haunt of Tree Pipit, 116; Pied Flycatcher observed, 133; Tawny Owl's nest, 295.

Fishwick, 303.

Fogo, Pyotknowes, 204; Corbie Hall, 242; Parish Rook Report, 228.

Foulden, xiv.; Meadow Pipit in winter, 114; Crawlea Plantations, 239; Parish Rook Report, 228.

Foulden West Mains, appearance of Sand Grouse, xx.; Reed-Bunting seen in February, 188; Mr. H. H. Craw's Report on Foulden Rookery, 228.

Fowl Carr, xxv.

German Ocean, xiii., xiv., xxiii.

Gibb's Cross at Wedderlie, a favourite haunt of Wild Geese, xxiii.

Girrick, Mr. Black's Report on Rooks in Parish of Neuthorn, 233.

Gladscleugh Burn, Gled, 301.

Gladswood, Kingfisher on Tweed, 267; name derived from Gled, 301.

Gledstane Forest, name derived from Gled, 301.

Godscroft or "*Gowkie*," derivation of name, 278.

Gordon, Great Grey Shrike killed, 124; Red-Backed Shrike seen, 127; rendezvous of Swallows before autumn migration, 138; Goldfinch, 150; Corn Bunting, 180; Parish Rook Report, 228; Crawlee Burn, 239.

Gordon Bog, Whinchat, 42; Stonechat, 45; Sedge Warbler, 77; Siskin, 153; Lesser Redpoll, 173; Reed-Bunting, 186.

Gordon Muir, Rev. Alexander Waugh's anecdote about Rooks, 235.

Grange Wood, Lesser Redpoll, 173; Barn Owl found dead in snow, 284; Short-Eared Owl frequently seen, 289.

Grantshouse, Chiffchaff, 69; Waxwing seen, 130; Kingfisher on the Eye, 269.

Great Dirrington Law, Ring Ouzel, 33.

Greenburn, 124; Rookery, 224.

Greencleugh Ridge, Ravens seen, 247.

Greenhead, Great Grey Shrike shot, 123; Hen-Harrier killed, 300.

Greenheugh, Rock Pipit, 118.
Greenhope, Stonechat, 45.
Greenknowe, Gordon, a rendezvous of Swallows before autumn migration, 138.
Greenlaw, Swallows assembling on roof of County Buildings, 138; Parish Rook Report, 229; Corbie Heugh Quarry and Raven, 242; Kingfisher on Blackadder, 270; Sea Eagle shot, 309.
Greenside Hill, arrival of Whinchat, 41.
Grueldykes, Great Grey Shrike killed, 123.
Gunsgreen, gathering-place of Martins before autumn migration, 143.

Hall Burn Wood Cleugh, Ring Ouzel's nest, 33.
Hallyburton, 309.
Hallydown, Waxwing seen, 130.
Hammerhall, occurrence of Grasshopper Warbler, 81.
Hardacres, occurrence of White's Thrush, 22.
Hardens, xiv.
Harelaw, Chirnside, haunt of Corn Bunting, 181.
Harelawside, 74.
Hutton, Sedge Warbler, 76; Parish Rook Report, 229, 230.
Harryburn, yellow-coloured Redbreast, 56; Mr. Romanes' Report on Rookeries in Parish of Lauder, see pp. 230, 231.
Hart Law, North, xxi.; *South*, xxi.
Hartlaw, North and South, Hartside, and Hindsidehill, names derived from Red Deer, xv.
Hatchednize, rookery, 224.
Haughhead, rookery, 225.
Hell's Cradle, Kingfishers' nest, 269.
Heriot Water, Pied Wagtail, 109; Sand Martin, 145.
Hernode, "venyson" reserved in lease of fifteenth century, xv.
Hirsel, introduction of the Pheasant, xx.; migration of Swallow, 138; rookery, 224; Barn Owl, 283.
Hog Hill, xxi.
Holy Island, Great Bustard shot, xvii.

Home Castle, xv.; James IV. visiting while on hawking expedition, xviii.; favourite resort of Swift, 256.
Hoolet Ha', 295.
Hoolets' Craig, Barn Owl, 281.
Horsley, Pied Wagtails' resort, 109; Sand Martin's nest, 145.
Houndwood, rookery, 223.
Howpark Burn, 239.
Hule Moss on Greenlaw Moor, a favourite haunt of Wild Geese, xxiii.; Sea Eagle shot, 309.
Hume, Parish Rook Report, 229.
Humehall, rookery, 229.
Hunny Crooks, old rookery, 228.
Hunt Law, "Titlin Cairn," 113.
Hunter's Well, Quicwood, Hen-Harrier, 300.
Huntington, Magpie, 203; Rough-Legged Buzzard trapped, 305; Sea Eagle, 308.
Huntlaw, xxi.; Sea Eagle seen, 308.
Hutton, Parish Rook Report, 229, 230.
Hutton Hall, Chiffchaff, 69; Tree Sparrow, 160; rookery, 229; Kingfisher, 268.
Hutton Hall Mill, 245.

Isle of May—migration reports—Song Thrush, 4; Blackbird, 27; Wheatear, 37; Redbreast, 51; Golden-Crested Wren, 66; Long-Tailed Titmouse, 89; Blue Titmouse, 98; Wren, 102; Pied Wagtail, 109; Meadow Pipit, 115; Rock Pipit, 118; Siskin, 152; Greenfinch, 154; House Sparrow, 159; Chaffinch, 163; Linnet, 170; Lesser Redpoll, 172; Bullfinch, 176; Corn Bunting, 181; Yellow Bunting, 183; Snow-Bunting, 189, 190; Starling, 195; Carrion Crow, 210; Hooded Crow, 214; Skylark, 252.

Jackdaws' Craig, 208.

Kames, rookery, 226.
Kelloe, rookery, 227.
Keppie Island, Cream-coloured Blackbird, 28.

TOPOGRAPHICAL INDEX. 329

Kilpallet Burn, 239.
Kimmerghame, bones and remains of Red Deer found, xv.; skull of Beaver discovered in Middlestots Bog, xvi.
Kimmerghame Mill, Swallow, 135.
Ladykirk, Parish Rook Report, 230; Kingfisher on Tweed, 267; Barn Owl, 281; Rough-legged Buzzard shot, 306; Sparrow-Hawk and young Pheasants, 311.
Ladykirk House, Mavis, 5.
Lady's Pocket Wood, Stonechat caught, 45.
Lamb Rig, xxi.
Lamberton, Tree Sparrow, 160; Snow-Bunting, 190; Skylark, 253; Hoopoe, 272.
Lamb's Mill, Hoopoe, 271.
Lamberton Moor, xiii.; Blackgame, xxii.; a favourite resort of the Dotterel, xxii., xxiii.; Ring Ouzel, 34; Raven, 247; Rough-Legged Buzzard, 306.
Lamberton Kirk, 240.
Lamberton Shiels, Rough-Legged Buzzard seen, 306.
Lamden Burn, Tree Pipit, 117.
Lamington Dean, Tree Pipit, 116.
Lammermuirs, bounding Berwickshire on the north, xiii.; Red Deer and Roe Deer in olden times, xv.; description of hills, xxi.; Dotterel formerly visiting in large flocks, xxii.; sloping to sea near Cockburnspath, xxiii.; Ring Ouzel's return in spring, 32; and its nest, 35; Stonechats' haunts, 45; Meadow Pipit, 113, 114; Rooks, 219; Raven, 248; Great Spotted Woodpecker, 262; Cuckoo, 274; Hen-Harrier, 298, 299; Earnscleuch Rig Hill, 310.
Lang Belt, *Mordington*, Nightjar sometimes observed, 260; Rough-Legged Buzzard seen, 306.
Langrig, Barn Owl, 284.
Langton, name of Raeclenghhead derived from Roe Deer, xv.; Redstart plentiful, 48; Wood Wren observed, 74; Siskin, 153; Parish Rook Report, 230; Raven, 241.
Langton Burn, 153.

Langton Tower, Raven, 241.
Lauder Common, Snow-Bunting, 191.
Lauder, xiii.; Meadow Pipit, 113; Great Grey Shrike, 124; Corn Bunting, 180; Magpie, 203; Parish Rook Report, 230, 231; Hen-Harrier, 299; Rough-Legged Buzzard, 305; Sea Eagle, 308, 310.
Lauder Barns, Waxwing, 130.
Lauderdale, Ring Ouzel, 34; Redstart, 48; Blackcap, 61; Tree Pipit, 117; Crossbill, 179; Carrion Crow, 211; Hooded Crow, 215; Raven, 247; Great Spotted Woodpecker, 262; Barn Owl, 282, 284; Hen-Harrier, 299.
Laughing Law, Ring Ouzel, 33.
Laverock Law Burn, 255.
Laverock Law, Coldingham, 255.
Laverock Law, Fogo, 255.
Laws, Honey Buzzard killed, 316.
Leader, xiii.; sources of, xxi.; Willow Wren, 71; Wren's song, 103; Siskin, 153; Lesser Redpoll, 173; Starling, 195; Jackdaw, 208; Kingfisher, 269; Barn Owl, 282.
Lees, rookery, 224; Kingfisher's nest, 267.
Leet, Buzzard shot on banks, 303.
Leetside, Barn Owl, 284.
Legerwood, a resort of wildfowl in olden times, xvii.; Dotterel in former days, xxii.; Brambling, 167; Bullfinch, 176; Magpie, 203; Parish Rook Report, 231; Barn Owl, 282.
Legerwood Loch, Goldfinch, 150.
Leitholm, Jackdaw, thieving propensities, 209.
Lennel, *Crowshiel*, 239.
Lennelhill, 297.
Liberties of Berwick, xiii.
Linkholm, Rock Pipit, 118.
Linthill, Redbreast's song, 52; Rooks prognosticating wind, 237; "Lady" Billie, 241.
Lippie Plantation, Great Spotted Woodpecker, 263.
Little Rooks, 239.
Lochton, Pied Blackbird, 29.
Longcroft Water, Ring Ouzel, 34; Kingfisher, 269.

Longformacus, Grouse formerly very scarce, xxii.; Whinchat, 42; Redstart, 48; Meadow Pipit, 113; Snow-Bunting, 190; Jackdaw's Craig, 208; Parish Rook Report, 232; Nightjar, 259; Great Spotted Woodpecker, 263; Sea Eagle, 308.
Longformacus House, Dipper's nest, 87; Starlings taking possession of dovecot, 194.
Lovers' Tryste, Long-Eared Owl, 288.
Lowrie's Knowes, 41.
Luggy, Siskin, 153; Lesser Redpoll, 173.
Lumsden Dean, Blackbird feasting on ivy berries, 29.
Lumsden Moss, haunt of Sedge Warbler, 77.
Lylestone, Brambling, 167.

Maines, rookery, 222.
Makerston, xiv.
Manderston, Goldfinch, 149; Superstition about Corbie Stone curing disease in cattle, 242.
Marchmont, effects of severe winter on Song Thrush, 5; migration of Swallow, 138; Starlings roosting in evergreens, 193; sex of Starlings roosting in evergreens during nesting season, 194; Jay, 201; Magpie — Pyotknowes, 204; Rookery Report, 220; Sea Eagle shot, 309.
Marygold, Twite, 174.
Meikle Law, xxi.
Mellerstain, Dotterel in former times, xxii.; Jackdaw stealing a will, 209; Rooks roosting at rookery at night in winter, 220; Rookery, 225; Kingfisher on lake, 270; Barn Owl's nest in chimney, 282; Buzzard, 302.
Merse, xv.; a resort of the Great Bustard in ancient times, xvii.; King James IV. visiting with his falconers, xviii.; surface transformed, xix.; formerly covered to a large extent with heather, xxi.; Blackbird, 27; Whinchat, 42; Garden Warbler, 63; Linnet, 168; Cuckoo, 274; Hen-Harrier, 296.

Mersington, rookery, 226.
Merton, xiv.; Sedge Warbler, 77; Goldfinch, 150; Siskin, 153; Magpie, 203; Parish Rook Report, 233; Nightjar, 260; Kingfisher on Tweed, 267.
Middlestots Boy, remains of Beaver found, xvi.
Millbank, Kingfisher on the Eye, 270.
Millburn, Rooks roosting in trees at night in winter, 220.
Millfield, Merton, Goldfinch, 150.
Milne Graden, Ring Onzel seen, 34; Golden-Crested Wren's nest, 66; Sedge Warbler, 77; Spotted Flycatcher nesting in tea-cup, 132; Jackdaw, 207; rookery, 224; Rooks prognosticating change in weather, 237; Great Spotted Woodpecker, 262; Wryneck seen, 265; Kingfisher on Tweed, 267; Barn Owl, 280, 284; severe winter of 1860-61, 283; Tawny Owl hooting during the day, 293.
Monynut, Grouse formerly very scarce, xxiii.; Great Spotted Woodpecker shot, 262.
Moorhouse, Laverock Law, 255.
Mordington, xiii., xiv.; Starlings roosting in evergreens, 193; Parish Rook Report, 232; Raven, 240; great flocks of Skylarks, 253; Great Spotted Woodpecker, 262; Hoopoe, 272.
Mordington Moor, Short-eared Owl, 289.
Muirside, Coldingham, Raven pecking out sheep's eyes, 246; Hen-Harrier's nest, 299.

Nabdean, Sedge Warbler, 76; Grasshopper Warbler heard, 79; Great Titmouse, 91; Lesser Redpoll, 173; Carrion Crow stealing young chickens, 211; depredations of Rook, 219; Buzzard, 303.
Nenthorn, Parish Rook Report, 233.
Netherbyres, Waxwing caught, 129; Raven, 247.
New Mills, Lauderdale, Brambling, 167.
Ninecairn Edge, xxi.

Ninewells, Redstart, 48; Wood Wren, 74; occurrence of Grasshopper Warbler, 81; rookery, 222; Kingfisher on Whitadder, 269.

Nisbet, Tawny Owl's nest, 295; Buzzard shot, 303.

Nisbet House, rookery, 227; Barn Owl, 284.

Nisbet Mill, Dipper's nest, 87.

Norham Castle, xiv.

Northfield, Raven, 246, 247.

North Hart Law, xxi.

Northumberland, xiv.

Norton Bridge, Lauderdale, Barn Owl, 282.

Nunlands, rookery, 227.

Oatleycleugh, Chiffchaff, 69; Wood Wren heard, 74; Buzzard seen, 303.

Old Bound Road, xiv.

Oldcambus, Mr. Hardy's notes on Song Thrush in frost, 3; —— on Missel Thrush, 8, 10, 12; —— on Redwing, 15; —— on Wheatear, 36, 37, 39; —— on Whinchat, 41, 42; —— on Sedge Warbler, 77; —— on Wren, 103; —— on Meadow Pipit, 114; occurrence of Waxwing, 129; Mr. Hardy's notes on Goldfinch, 148, 149; —— on Brambling, 167; —— on Corn Bunting, 181; —— on Yellow Bunting, 184; —— on Reed-Bunting, 187; —— on Snow-Bunting, 190; —— on Raven, 245, 247; —— on Nightjar, 259; Sea Eagle occasionally shot, 308.

Oldcambus Dean, Fieldfares roosting in furze, 20; arrival of Ring Ouzel, 32; Stonechat in winter, 44; a breeding place of Stonechat, 45.

Oldcastles, Corn Bunting, 181.

Oldhamstocks, Parish Rook Report, 233.

Old Heronry Wood, Paxton, Chiffchaff, 68; Grey Wagtail, 111; nest of Sparrow-Hawk, 314.

Old Thirlstane, Goldfinch, 150; Lesser Redpoll, 173.

Ordweil, a favourite resort of Cuckoo, 274.

Outlawhill, Hen Harrier breeding, 299.

Oxendean, occurrence of Red-backed shrike, 127; Barrowmill woods, 222.

Oxton, Brambling in winter, 167.

Paddock Cleugh, Sand Martin, 145.

Paxton, xiv.; appearance of Stock Dove, xx.; Mavis singing, 2; Missel Thrush's nest, 11; favourite resort of Redwing, 16; Blackbird's song, 30; Redstart's first appearance, 48; its nest in pump, 49; Blackcaps, 60, 62; Garden Warbler, 63; Wood Wren, 74; Sedge Warbler, 77; Great Titmouse destroying bees, 92; Marsh Titmouse, 96; Wrens frozen to death, 103; Pied Wagtail's nest, 109; Grey Wagtail, 111; Tree Pipit, 116; Swallow, 136, 137; Sand Martin, 144, 145; Siskin, 153; Brambling, 167; Lesser Redpoll, 173; Bullfinch, 176; Crossbill, 179; Corn Bunting seldom seen, 180; Yellow Bunting's nest, 185; Reed-Bunting, 187; Magpie, 203; Carrion Crow, 210, 212; Rook, 219, 220; Rookery, 229; Barn Owl, 281, 284; Hen-Harrier seen, 300; Buzzard observed, 303; Sparrow-Hawk's nest, 314.

Paxton House, Missel Thrush, 13; Dipper, 87; Pied Flycatcher, 134; Tree Creeper, 147; Starling, 193; Great Spotted Woodpecker, 262; Long-eared Owl, 287.

Paxton South Mains, Skylark, 253.

Paxton Toll, Stonechat, 46.

Pease Bridge, Whinchat, 42; Marsh Titmouse's nest, 95; Tree Sparrow, 160, Buzzard, 302.

Pease Burn, Thrush, 5; Ring Ouzel, 33; Grey Wagtail, 111.

Pease Dean, Blackbirds feasting on ivy berries, 29; Blackcap, 60, 61; Chiffchaff, 68; Wood Wren, 73; Grasshopper Warbler, 81; Marsh Titmouse, 95; White Wagtail said to have been observed, 110; Tree Pipit, 116; Barn Owl, 282.

Pease Mill, Sedge Warbler, 77; Siskin, 153.
Peat Law, 301.
Peelwalls, rookery, 221.
Peely Braes, Rooks roosting on trees at night in winter, 222, 225.
Penmanshiel, Ring Ouzel, 33; Sedge Warbler, 77; Grasshopper Warbler, 81; Waxwing, 130; Goldfinch, 149; House Sparrow, 159; Brambling, 167; Lesser Redpoll, 173; Corn Bunting, 181; Yellow Bunting, 184; Reed-Bunting, 187; Snow-Bunting, 190, 191; Nightjar, 258; Hen-Harrier, 299, 300.
Penmanshiel Moor, Whinchat, 42; Stonechat, 45.
Penmanshiel Wood, Tree Pipit, 116; Jay, 200; Magpie, 203; Nightjar, 259; Great Spotted Woodpecker, 263; Honey Buzzard, 315, 316.
Petticowick, xxv.; Chough, 199; Raven, 245.
Pilmore, Lauderdale, Brambling, 167.
Piperton Hill, Stonechat, 45; old name, 310.
Pistol Plantation, ash-coloured Blackbird, 29; Chiffchaff, 69; Wood Wren, 74; Long-Tailed Titmouse, 89; Pied Flycatcher, 134; Buzzard, 303.
Polwarth, Parish Rook Report, 233, 234.
Polwarth Kirk, Barn Owl, 283.
Press, Crossbill, 179; Nightjar, 260.
Preston, Grasshopper Warbler, 80; Goldfinch, 150.
Preston Bridge, Kingfisher, 269.
Primrosehill, Buzzard, 303.
Purveshall, rookery, 226.
Pyatshaw, 204.
Pyatshaw Burn, 204.
Pyatshaw Knowes, 204.
Pyatshaw Ridge, 204.
Pyotknowes, 204.

Quixwood, appearance of Sand Grouse, xx.; Whinchat, 42; Sand Martin, 145; Great Spotted Woodpecker, 263; Cuckoo, 278; Hen-Harrier, 300; Buzzard, 306.

Raecleughhead, name derived from Roe Deer, xv.
Rammel Cove, Raven, 245.
Rampart at St. Abb's Head, xxv.
Ravelaw, Barn Owl, 284.
Raven's Brae, 242.
Raven's Craig, 242, 243.
Raven's Heugh, xxv., 242.
Raven's Knowe, 242.
Rawburn, name derived from Roe Deer, xv.; Hooded Crow, 214.
Red Clews Cleugh, Nightjar, 259.
Redheugh, a favourite resort of the Dotterel, xxii.; Song Thrush in flocks, 3; Whinchat, 42; Sedge Warbler, 77; Sand Martin, 145; Reed-Bunting, 187; Chough, 197.
Redheugh Hill, Whinchat, 42, 43; Stonechat on heights, 45.
Red Cot Waas, Wheatear, 40.
Renton, Pied Wagtail, 109; Kingfisher on the Eye, 269.
Reston, Great Grey Shrike, 123; Tree Sparrow, 160.
Retreat, Fieldfare, 19; Redstart, 48; Tree Pipit, 117; Jay, 201; Great Spotted Woodpecker, 263; Buzzard, 303.
Rhymer's Mill, Earlston, Lesser Redpoll, 173.
Rigfoot, Whinchat, 42.
Roe Deer, xv.
Rooks—and *Little Rooks*—rocks off coast of Dowlaw, 239.
Rosy Bank, Coldstream, effects of severe winter of 1878-79 on the Song Thrush, 5.
Roxburgh, xiv.
Rumbleton Law, old rookery, 228; Hen-Harrier shot, 300.

St. Abb's Head, xiii; Redwing in severe weather, 15; Stonechat, 45; Martin nesting on rocks, 142; Tree Sparrows' nests, 160; Linnets' nests, 170; former haunts of Chough, 197, 198, 199; Jackdaw, 207; Raven breeding, 243; harrying Ravens' nests, 244, 245, 248; Raven pecking out sheep's eyes, 246, 247; Sea Eagle observed, 308; Earnsheugh, 310.

St. Helen's Kirk, Wheatear singing, 40; Rock Pipit, 118; curiosity of Jackdaw, 208.
St. Leonards, rookery, 231.
Saltpan-hall, Sand Martin's nest, 145.
Scarsheugh, Barn Owl's nest in former times, 280, 281; Tawny Owl, 295.
Scenes Law, xxi.
Siccar Point, Starling nesting in rocks, 195; Raven seen, 248.
Skaithmuir, Barn Owl seen in Dec. 1887, 284.
Skelly, xxv.
Soldier's Dyke, haunt of Whinchat, 42.
South Fallaknowe, Sea Eagle trapped, 308.
South Hart Law, xxi.
Spital House, cock-nests of Wren, 105; Reed-Bunting's nest, 188; Great Spotted Woodpecker shot, 262; Tawny Owl's nesting-place, 295.
Spottiswoode, introduction of the Pheasant, xx.; Blackgame numerous, xxii.; Ring Ouzel, 33; Brambling, 167; nesting of Lesser Redpoll, 173; Bullfinch, 176; Snow-Bunting, 190; Hooded Crow, 215; rookery, 234; weather lore about Rook, 237; Barn Owl, 282.
Stainrig, Wood Wren, 74; Waxwing shot, 129; rookery, 226.
Staneshill, Barn Owl, 281.
Starchhouse Toll, Hoopoe, 271.
Stockbridge, Pied Wagtail, 109; Sand Martin, 145.
Stonymoor, *Duns*, Corn Bunting formerly plentiful, 180.
Smailholm, xiv.
Stitchel, xiv.
Sunnyside, *Anchencrow*, Magpie's nest, 203.
Sunnyside, *Milne Graden*, Sand Martin's breeding place, 145.
Sunwick, Long-Eared Owl, 288.
Swallow Craig, Martin's nest, 142; derivation of name, 143; old breeding quarters of the Raven, 245.
Swallow-dean, derivation of name, 143.

Swallowheugh, derivation of name, 143.
Swinton, a resort of the Wild Boar in ancient times, xvi.; Great Grey Shrike, 124; White Chaffinch, 164; Parish Rook Report, 234; Earnslaw, 310.

Tarf Law, xxi.
Templehall, rookery, 223.
Thirlestane Castle, introduction of the Pheasant, xx.; Redstart's nest, 48; Blackcap, 61; Crossbill in fir wood, 179; rookery, 231; Barn Owl, 282.
Thornydykes, Goldfinch observed, 150.
Threeburn Ford, Lauderdale, Brambling feeding on lint, 167.
Threeburn Grange, Great Spotted Woodpecker shot, 263.
Threepwood, Crane shot, xvii.
Thrummycar Heugh, a former haunt of the Chough, 198.
Titlin Cairn, Meadow Pipit, 113.
Todrig, rookery, 229.
Trabroun, Snow-Bunting in flocks, 190.
Tweed, xiii., xiv.; Barn Owl nesting in precipices, xx.; Cream-coloured Blackbird at Milne Graden, 28; Stonechat in autumn, 46; Sedge Warbler at Paxton, 77; Dipper, 87; Pied Wagtail's nesting place, 109; Pied Flycatcher seen, 133; Sand Martin, 145; Bullfinch, 176; Starling nesting in rocks overhanging stream, 195; Jackdaws' haunts, 207; Carrion Crow's nest, 212; Paxton rookery, 219, 220; Crowshiel and Crow Green, 239; Great Spotted Woodpecker, 262, 263; Kingfisher, 267; Barn Owl formerly breeding in rocks at Scarsheugh, 280, 281; nesting places now occupied by Jackdaws, 283; Longeared Owl, 287; Tawny Owl hooting, 292, 293; at Scarsheugh, 295; Hen-Harrier seen, 300; Gladswood, 301.
Twinlawford, Craw Cleugh, 239.

Upper Moorhouse, Wheatear, 37.

TOPOGRAPHICAL INDEX.

Waddel's Cairn, xxi.
Watch Water, Ring Ouzel, 33; Redstart, 48; Twite, 175; Kingfisher, 270.
Wedderburn, Sir David Home taking Partridges with Falcons, xviii.; Wood Wren, 74; Bullfinch, 176.
Wedderlairs, xxi.; Meadow Pipit, 113.
Wedder Law, xxi.
Wether Law, xxi.
Wedderlie, Wheatear, 37; Cuckoo, 274.
Weir Burn, Ring Ouzel, 33.
West Greenfield, Twite, 175.
West Hope, Raven's nest, 245.
West Hurker, xxv.
Weston Thirl Cliff, Raven, 246.
Westruther, remains of Red Deer found in parish, xv.; Ring Ouzel, 33; Wheatear, 37; Bullfinch, 176; Snow-Bunting, 190; Magpie, 203, 204; Hooded Crow, 215; Parish Rook Report, 234; Tawny Owl, 295.
Whalpland Burn, Gled, 301.
Wheelburn, Hen-Harrier's nest, 299.
Whitadder, Barn Owl nesting in precipices, xx.; sources of, xxi.; Whitethroat's song, 58; Garden Warbler, 63; Willow Wren, 71; Dipper, 86, 87; Wren, 103; Grey Wagtail, 111, 112; Martin, 142; Sand Martin, 144, 145; Goldfinch, 150; Reed-Bunting, 186; Starling, 195; Jackdaw, 208;

Carrion Crow, 212; "Raven's Knowe," 242; "Blue Braes," 245; Raven's nest in precipices, 248; Kingfisher, 267, 268, 269; Barn Owl, 281, 283; Hen-Harrier, 300; Sparrow-Hawk, 313.
Whitchester, Wheatear's nest, 37; Whinchat, 42; Stonechat, 45; Twite seen, 175; Reed-Bunting, 187; Raven seen, 248.
Whiteburn, remains of Red Deer found, xv.; Reed-Bunting, 187.
Whitecross, Crossbill, 179.
White Gate, Blackerston, Great Spotted Woodpecker, 263.
Whinny Park, Hen-Harrier's nest, 299.
Whitehall, Blackcap, 60; Garden Warbler, 63; Grasshopper Warbler, 81; rookery, 222.
Whiterig, Reed-Buntings seen in December, 187; rookery, 221.
Whitfield, Crossbill, 179.
Whitrig Bog, remains of Red Deer found, xv.
Whitslaid, Starling, 195.
Whitsome, Parish Rook Report, 234.
Whitsome Village, Swallow building in "lums," 136; Eaglehall, 310.
Willie's Law, xxi.
Windshiel, Whinchat, 42.
Winfield, superstition about Magpie, 204; Rooks pulling up young turnips, 218.
Witches' Knowe, Raven, 240.
Wrink Law, Ring Ouzel, 33.

Printed by T. & A. CONSTABLE, Printers to Her Majesty,
at the Edinburgh University Press.

BOOKS

ON

SPORT & NATURAL HISTORY

PUBLISHED BY

DAVID DOUGLAS

Small Folio, price £21, with Sketches of Scenery and Animal Life by some of the best British and American Artists and Etchers.

THE
RISTIGOUCHE
AND ITS
SALMON FISHING
WITH A CHAPTER ON ANGLING LITERATURE
By DEAN SAGE

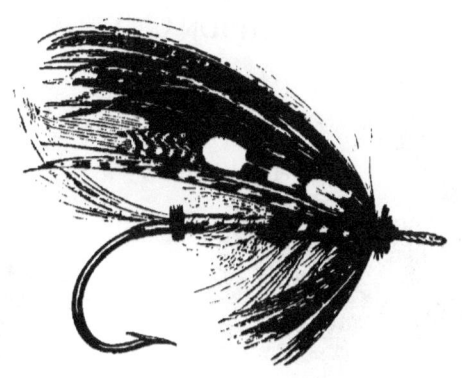

EDINBURGH: DAVID DOUGLAS
1888

Only 105 Copies printed.

One Volume, Small 4to. 24s.

Also a Cheaper Edition, with Lithographic Illustrations,
Demy 8vo. 12s.

Wild Men & Wild Beasts
Scenes in Camp & Jungle
By Lt. Col. Gordon Cumming
Illustrated by Col. R. Baigrie and others.

EDINBURGH: DAVID DOUGLAS. MDCCCLXXI.

One Volume, Royal 8vo. 50s.

WITH 40 FULL-PAGE ILLUSTRATIONS OF SCENERY AND ANIMAL LIFE, DRAWN BY
GEORGE REID, R.S.A., AND J. WYCLIFFE TAYLOR, AND ENGRAVED BY AMAND DURAND.

NATURAL HISTORY & SPORT

IN MORAY

By CHARLES ST. JOHN
AUTHOR OF "WILD SPORTS IN THE HIGHLANDS"

EDINBURGH: DAVID DOUGLAS
1882

Two Volumes, Crown 8vo, Illustrated. 21s.

A TOUR IN SUTHERLANDSHIRE

WITH EXTRACTS FROM THE FIELD BOOKS OF A SPORTSMAN AND NATURALIST

By CHARLES ST. JOHN

AUTHOR OF "NATURAL HISTORY AND SPORT IN MORAY"

SECOND EDITION

WITH AN APPENDIX ON THE FAUNA OF SUTHERLAND

BY J. A. HARVIE-BROWN AND T. E. BUCKLEY

EDINBURGH: DAVID DOUGLAS
1884

One Volume, Demy 8vo, with Maps and Illustrations.   12s.

NOTES AND SKETCHES

FROM THE

WILD COASTS OF NIPON

WITH CHAPTERS ON CRUISING AFTER PIRATES
IN CHINESE WATERS

By CAPTAIN H. C. ST. JOHN, R.N.

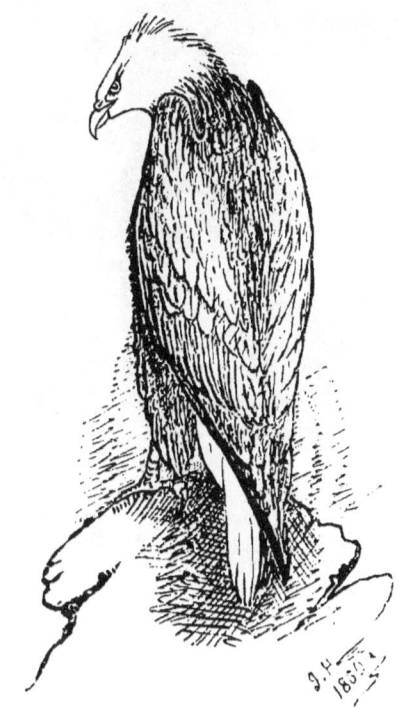

EDINBURGH: DAVID DOUGLAS
1880

One Volume, Demy 8vo. 18s.

SASKATCHEWAN

AND

THE ROCKY MOUNTAINS

*A DIARY AND NARRATIVE OF TRAVEL, SPORT, AND ADVENTURE DURING
A JOURNEY THROUGH THE HUDSON BAY COMPANY'S TERRITORIES*

By THE EARL OF SOUTHESK, K.T.

WITH MAPS AND ILLUSTRATIONS

EDINBURGH: DAVID DOUGLAS

One Volume, Demy 8vo, with Etchings and Map. 8s. 6d.

THE
CAPERCAILLIE IN SCOTLAND

WITH SOME ACCOUNT OF THE EXTENSION OF ITS RANGE SINCE ITS RESTORATION AT TAYMOUTH IN 1837 AND 1838

By J. A. HARVIE-BROWN, F.Z.S.

MEMBER OF THE BRITISH ORNITHOLOGISTS' UNION, ETC.

EDINBURGH: DAVID DOUGLAS. MDCCCLXXIX

*In One Volume, Small 4to, with Map and Illustrations by Messrs.
J. G. Millais, T. G. Keulemans, Samuel Read, and others.* 30s.

A
VERTEBRATE
FAUNA
OF
SUTHERLAND, CAITHNESS
AND
WEST CROMARTY

By J. A. HARVIE-BROWN, F.R.S.E., F.Z.S.
AND
T. E. BUCKLEY, B.A., F.Z.S.

EDINBURGH
DAVID DOUGLAS

Nearly Ready. In Two Volumes, Demy 8vo. To Subscribers only.
Profusely Illustrated with Etchings and Lithographs.

THE
BIRDS OF BERWICKSHIRE

WITH REMARKS ON THEIR LOCAL DISTRIBUTION
MIGRATION, AND HABITS, AND ALSO ON THE
FOLK-LORE, PROVERBS, POPULAR RHYMES
AND SAYINGS CONNECTED WITH THEM

BY

GEORGE MUIRHEAD, F.R.S.E., F.Z.S.

MEMBER OF THE BRITISH ORNITHOLOGISTS' UNION, MEMBER OF THE
BERWICKSHIRE NATURALISTS' CLUB, ETC.

EDINBURGH: DAVID DOUGLAS
1889

One Volume, Demy 8vo. 15s.

THE ART OF GOLF

By SIR WALTER SIMPSON, Bart.

With 20 Illustrations from Instantaneous Photographs of
Professional Players, chiefly by A. F. Macfie, Esq.

EDINBURGH: DAVID DOUGLAS

In the Press, One Volume, Demy 8vo, and Large Paper Edition, with additional Illustrations, Small 4to.

A HISTORY OF CURLING

SCOTLAND'S AIN GAME

AND OF FIFTY YEARS OF
THE ROYAL CALEDONIAN CURLING CLUB

Edited by

The Rev. JOHN KERR, M.A., Dirleton

The Spirit of Curling

EDINBURGH: DAVID DOUGLAS
1889

A TREATISE ON ANGLING

HOW TO CATCH TROUT

By THREE ANGLERS.

Illustrated. Price 1s., by Post, 1s. 2d.

The aim of this book is to give within the smallest space possible such practical information and advice as will enable the beginner without further instruction to attain moderate proficiency in the use of every legitimate lure.

"A delightful little book, and one of great value to anglers."—*Scotsman.*
"The advice given . . . is always sound."—*Field.*
"As perfect a compendium of the subject as can be compressed within eighty-three pages of easily read matter."—*Scotch Waters.*
"A well written and thoroughly practical little work."—*Land and Water.*
"The most practical and instructive work of its kind in the literature of angling."—*Dundee Advertiser.*

"*A Delightful Guide for a Country Ramble.*"

A YEAR IN THE FIELDS

By JOHN WATSON.

Fcap. 8vo. Price 1s., by Post, 1s. 2d.

"A charming little work: a lover of life in the open air will read the book with unqualified pleasure."—*Scotsman.*
"A brief but prettily written account of the natural phenomena incident to each month."—*Liverpool Mercury.*

ALEX. PORTER.

THE GAMEKEEPER'S MANUAL

BEING AN EPITOME OF THE GAME LAWS OF ENGLAND AND SCOTLAND, AND OF THE GUN LICENCES AND WILD BIRDS ACTS

FOR THE USE OF GAMEKEEPERS AND OTHERS INTERESTED
IN THE PRESERVATION OF GAME

By ALEXANDER PORTER, CHIEF CONSTABLE OF ROXBURGHSHIRE

Second Edition. Crown 8vo, Price 3s., Post free.

"A concise and valuable epitome to the Game Laws specially addressed to those who are engaged in protecting game."—*Scotsman.*
"Quite a store-house of useful information. . . . Although not pretending to be a 'law book,' this work will certainly serve the purpose of one; no subject being omitted that comes within the province of the game laws."—*Glasgow Herald.*

ROBERT MORETON.

ON HORSE-BREAKING

By ROBERT MORETON

Second Edition. One Volume, Crown 8vo. Price 1s.

EDINBURGH: DAVID DOUGLAS

In One Volume, Small 4to, with Maps, and Illustrated by Etchings, Cuts, Lithographs and Photogravure plates. 30s.

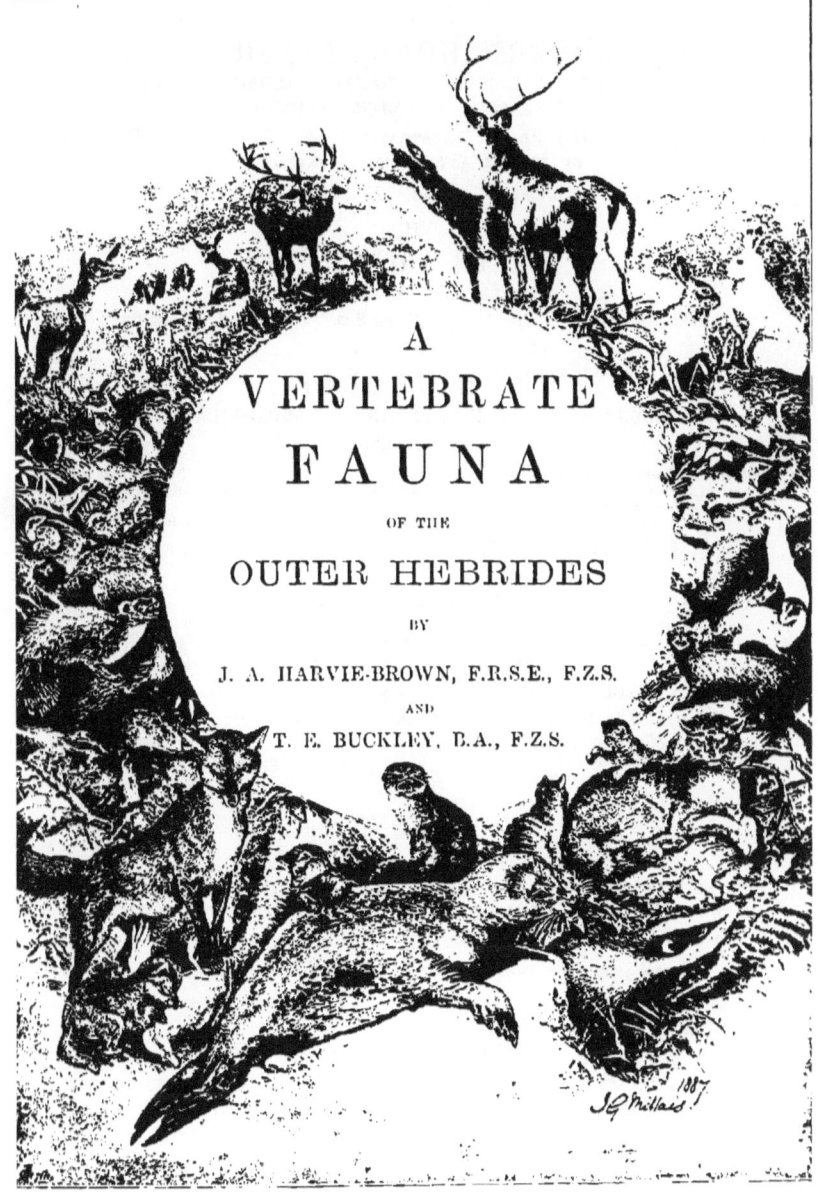

A VERTEBRATE FAUNA
OF THE
OUTER HEBRIDES

BY

J. A. HARVIE-BROWN, F.R.S.E., F.Z.S.

AND

T. E. BUCKLEY, B.A., F.Z.S.

EDINBURGH: DAVID DOUGLAS, CASTLE STREET

E. L. ANDERSON'S
BOOKS ON HORSEMANSHIP.

I.
MODERN HORSEMANSHIP
A NEW METHOD OF TEACHING RIDING AND TRAINING BY MEANS OF INSTANTANEOUS PHOTOGRAPHS FROM THE LIFE.

THIRD EDITION, WITH FRESH ILLUSTRATIONS OF "THE GALLOP CHANGE," OF UNIQUE AND PECULIAR INTEREST.

In One Volume, Demy 8vo. Illustrated. Price 21s.

II.
VICE IN THE HORSE
AND OTHER PAPERS ON HORSES AND RIDING.

New Edition, with Additions. Demy 8vo. Illustrated. Price 5s.

III.
THE GALLOP
ILLUSTRATED BY INSTANTANEOUS PHOTOGRAPHY.

Small 4to. 2s. 6d.

"It is impossible to read a page in any of his books without recognising the fact that this is a practical horseman speaking from long experience of an art which he has devotedly studied and practised."—*The Saturday Review.*

"The best new English work on riding and training that we can recommend is the book 'Modern Horsemanship.'"—*The Sport Zeitung, Vienna.*

"Every page shows the author to be a complete master of his subject."—*The Field.*

COLONEL DODGE.
A CHAT IN THE SADDLE
OR, PATROCLUS AND PENELOPE

BY THEO. A. DODGE, LIEUTENANT-COLONEL, U.S. ARMY.

Illustrated by 14 Instantaneous Photographs. Demy 8vo. Price 21s.

"A very learned and charming book about horses."—*Graphic.*

"We recommend Col. Dodge's work as one of the most important and valuable treatises upon the art of riding that we have in our language."—*Saturday Review.*

COLONEL CAMPBELL.
MY INDIAN JOURNAL
CONTAINING DESCRIPTIONS OF THE PRINCIPAL FIELD SPORTS OF INDIA, WITH NOTES ON THE NATURAL HISTORY AND HABITS OF THE WILD ANIMALS OF THE COUNTRY

BY COLONEL WALTER CAMPBELL, AUTHOR OF "The Old Forest Ranger."

Small Demy 8vo, with Drawings on Stone by Wolf. Price 16s.

GENERAL GORDON.
THE ROOF OF THE WORLD
BEING THE NARRATIVE OF A JOURNEY OVER THE HIGH PLATEAU OF TIBET TO THE RUSSIAN FRONTIER, AND THE OXUS SOURCES ON PAMIR.

BY MAJOR-GENERAL T. E. GORDON, C.S.I.

With Numerous Illustrations. Royal 8vo. Price 31s. 6d.

EDINBURGH: DAVID DOUGLAS.

www.ingramcontent.com/pod-product-compliance
Lightning Source LLC
Chambersburg PA
CBHW022332230426
43664CB00040B/408